SpringerBriefs in Education

We are delighted to announce SpringerBriefs in Education, an innovative product type that combines elements of both journals and books. Briefs present concise summaries of cutting-edge research and practical applications in education. Featuring compact volumes of 50 to 125 pages, the SpringerBriefs in Education allow authors to present their ideas and readers to absorb them with a minimal time investment. Briefs are published as part of Springer's eBook Collection. In addition, Briefs are available for individual print and electronic purchase.

SpringerBriefs in Education cover a broad range of educational fields such as: Science Education, Higher Education, Educational Psychology, Assessment & Evaluation, Language Education, Mathematics Education, Educational Technology, Medical Education and Educational Policy.

SpringerBriefs typically offer an outlet for:

- An introduction to a (sub)field in education summarizing and giving an overview of theories, issues, core concepts and/or key literature in a particular field
- A timely report of state-of-the art analytical techniques and instruments in the field of educational research
- A presentation of core educational concepts
- An overview of a testing and evaluation method
- A snapshot of a hot or emerging topic or policy change
- An in-depth case study
- A literature review
- A report/review study of a survey
- An elaborated thesis

Both solicited and unsolicited manuscripts are considered for publication in the SpringerBriefs in Education series. Potential authors are warmly invited to complete and submit the Briefs Author Proposal form. All projects will be submitted to editorial review by editorial advisors.

SpringerBriefs are characterized by expedited production schedules with the aim for publication 8 to 12 weeks after acceptance and fast, global electronic dissemination through our online platform SpringerLink. The standard concise author contracts guarantee that:

- an individual ISBN is assigned to each manuscript
- each manuscript is copyrighted in the name of the author
- the author retains the right to post the pre-publication version on his/her website or that of his/her institution

More information about this series at http://www.springer.com/series/8914

Geraldo W. Rocha Fernandes
António M. Rodrigues
Carlos Alberto Rosa Ferreira

Using ICT in Inquiry-Based Science Education

Geraldo W. Rocha Fernandes
Departamento de Ciências Biológicas
Universidade Federal dos Vales do
Jequitinhonha e Mucuri
Diamantina, Minas Gerais, Brazil

Carlos Alberto Rosa Ferreira
UIDEF, Instituto de Educação
Universidade de Lisboa, Faculdade de
Motricidade Humana
Lisboa, Portugal

António M. Rodrigues
LaPed, Laboratório de Pedagogia
Faculdade de Motricidade Humana
UIDEF, Instituto de Educação
Universidade de Lisboa
Lisboa, Portugal

ISSN 2211-1921   ISSN 2211-193X (electronic)
SpringerBriefs in Education
ISBN 978-3-030-17894-9   ISBN 978-3-030-17895-6 (eBook)
https://doi.org/10.1007/978-3-030-17895-6

This Springer imprint is published by the registered company Springer Nature Switzerland AG.
The registered company address is: Gewerbestrasse 11, 6330 Cham, Switzerland

Preface

This book presents the main Information and Communication Technologies (ICTs) and their effects on science education. It characterises the main theoretical approaches that support the use of ICTs. It also presents an analysis of inquiry-based science education and the possibility of employing ICT as a supporting tool when performing inquiry activities in science teaching.

Considering the overall organisation of the present book, structured to science teachers and public higher education institution, three distinct chapters were articulated. In particular:

- Chapter 1: This chapter analyses the main ICTs and their effects on science education. Different studies were used to reflect on two aspects: (1) science teaching and learning through ICT tools and digital resources and (2) science teaching and learning through the use of digital laboratories. There are different ICT and digital resources used in science education: software, Internet, computer games, and virtual and remote laboratories, among others. In this chapter examples of the effective use of ICT are included; however, studies suggest that the essential component for ICT-based science teaching and learning is determined by the teacher's pedagogical approaches. In particular, we determined that there are few studies that report strategies and didactics for the use of ICT in science classes, and they place a greater emphasis on outcomes and resources rather than on the process.

- Chapter 2: This chapter presents an analysis of the main theoretical approaches on science education mediated by ICT. Different studies were used to reflect on four aspects: (1) Approaches to teaching and learning through the use of ICT (reflecting trends in a "theory of technological education" or "cognitive tools"); (2) Cognitivist approaches (with emphasis on "social constructivism and sociocultural theory", "constructivist approaches" and "the effects of collaborative work" allowed by the use of ICT); (3) Approaches based on inquiry, research, projects and case studies (with a tendency towards renewal of the science curriculum); and (4) Approaches that emphasise conceptual knowledge (investigating "conceptual understanding" and "conceptual change" facilitated by the use of ICT). The main contribution of this chapter is the presentation of some theoretical studies and reflections for the use of ICTs in science education. The use of ICT is still planned and supported by traditional theoretical trends in teaching, learning, knowledge and curriculum design.

- Chapter 3: This chapter presents an analysis of inquiry-based science education and the possibility of employing ICT as a supporting tool when performing inquiry activities in science teaching. Different studies were used to reflect on two aspects: (1) the main ICTs that are used in inquiry activities and (2) the main steps in inquiry activities that are used in science education and their approaches to the use of ICT. This chapter discusses that, together with teacher supervision, the use of ICT for developing inquiry-based science education allows students to develop more active work styles, improved attitudes towards science, better conceptual and theoretical understanding, improved reasoning, better modelling capabilities and improved teamwork, along with improvements in other abilities.

We hope that this book will prove to be a functional scaffold for effectively approaching science education and use of ICTs, integrating inquiry-based science education and the use of ICTs. Let the journey begin!

Diamantina, Brazil   Geraldo W. Rocha Fernandes
Lisboa, Portugal   António M. Rodrigues
Lisboa, Portugal   Carlos Alberto Rosa Ferreira

Acknowledgements

The authors would like to express their gratitude to the members of the project of social and digital inclusion "Space, Challenges and Opportunities", to the Foundation for Science and Technology of Portugal and to the professors of the Faculty of Human Kinetics, University of Lisbon (Portugal), for their constructive discussions, opinions and comments on the main issues addressed in this book.

Contents

Chapter 1
ICT-Based Science Education: Main Digital Resources and Characterisation

1.1 Introduction

Although educational policies and guidelines have put enormous effort into positioning Information and Communication Technologies (ICT) as a central tenet of contemporary education, their use in educational settings still encounters resistance from many teachers (Athanassios, 2010). Even though access to computers has increased in schools, in most cases, teachers continue to use ICT primarily for formal academic tasks (to obtain information from the Internet) or administrative purposes (to develop lesson plans, worksheets and assessment tests) and not as a tool to support students in active learning (Chang & Tsai, 2005; Dori & Belcher, 2005).

Despite the difficulties faced in the introduction of ICT into educational settings, some research has been conducted to understand and present the general trends in science education mediated by ICT for Education (ICTE). For example, the review by Bell and Bell (2003) presents a bibliography of over 50 articles published between 1994 and 2003 on the use of ICT in K-12 science teaching. The review by Lee et al. (2011) is one of the few articles discussing trends in computer and Internet advancements, as well as their applications to science education. However, only a small number of articles characterise the teacher's teaching process and how the student's science learning occurs.

Because there is an existing scenario that seeks to integrate different ICT into educational settings, this chapter analyses *the main digital resources used and their effects* on science teaching. To further this objective, we organised our reflection based on two questions:

1. What are the main digital technologies (ICT tools and digital resources) used to teach science in formal educational settings?
2. How can science teaching and learning mediated by ICT tools and digital resources be characterised?

© The Author(s), under exclusive license to Springer Nature Switzerland AG 2019
G. W. Rocha Fernandes et al., *Using ICT in Inquiry-Based Science Education*,
SpringerBriefs in Education, https://doi.org/10.1007/978-3-030-17895-6_1

For a better understanding of this chapter, we included studies that explore the resources and processes that enable ICT-based science teaching and learning within formal teaching spaces.

1.2   Science Education and Digital Technologies

Although educational policies and guidelines have put enormous effort into positioning ICTs as a central subject of contemporary education, their use in educational settings still encounters resistance from many teachers (Athanassios, 2010). Even though access to computers has increased in schools, in most cases, teachers continue to use ICT primarily for formal academic tasks (to obtain information from the Internet) or administrative purposes (to develop lesson plans, worksheets and assessment tests) and not as a tool to support students in active learning (Chang & Tsai, 2005; Dori & Belcher, 2005).

Charlier, Peraya, and Collectif (2007) use the term ICTE, that is, Information and Communication Technologies for Education (ICTE), which include the different digital tools that can be used in education and teaching (ICTE = ICT + Education). Strømme and Furberg (2015) use the term digital resources to characterise the tools that are embedded in computer-based inquiry environments and that could support student learning. Examples of digital resources are dynamic or static visualisations, computer simulations, interactive tasks, collaboration- and argumentation-supporting tools, domain-specific text, etc., designed to represent a scientific phenomenon and/or central scientific concept.

For the current book, we will use the term digital technologies (ICTE tools and digital resources) to characterise the studies in ICT-based science education. For us, ICTE tools are a hardware perspective, while the digital resources are from the perspective of digital content. Examples of ICTE tools that we could find in recent studies are interactive whiteboard (IWB) (Warwick, Mercer, Kershner, & Staarman, 2010), mobile learning environments (MLE) (Ekanayake and Wishart 2015; Price et al. 2013), Moodle platform (Pombo et al. 2012), computers/laptops (Howard et al. 2015; Nielsen et al. 2014; Şad and Göktaş 2014), etc. Examples of digital resources are simulation (Anastopoulou et al. 2011; Khan, 2010; Lindgren & Schwartz, 2009; Plass et al., 2012), the Internet and Web (Gelbart, Brill, & Yarden, 2009; Katz, 2011; Lee et al., 2011; She et al., 2012), multimedia and hypermedia (Starbek, Starčič Erjavec, & Peklaj, 2010; Tolentino et al., 2009; Zheng, Yang, Garcia, & McCadden, 2008), animation (Barak, Ashkar, & Dori, 2011; Dalacosta, Kamariotaki-Paparrigopoulou, Palyvos, & Spyrellis, 2009), game (Squire & Jan, 2007), wiki (Chen et al. 2015; Donnelly and Boniface 2013; Kim, Miller, Herbert, Pedersen, & Loving, 2012), educational software (Lavonen et al. 2003; Valtonen et al. 2013), movies/video (Ling Wong et al. 2006; Roth et al. 2011), videoconference (McConnell et al. 2012), etc.

Teachers' PD is seen as the most important aspect of digital technology integration (ICTE tools and digital resources), and it has been repeatedly identified as a top priority in education policies, i.e. one of the PD goals for teachers is to be familiar with the emerging issues in digital technology integration (Hsu, 2006). The number of studies on the effectiveness of technologies and how to introduce them into the science curriculum or in PD programs has been increasing annually; however, little is known about its use in the classroom or about its relation with science teacher's PD.

1.3   What Are the Main Digital Technologies (ICTE Tools and Digital Resources), and How Are They Used to Teach Science?

Science education addresses an organised body of knowledge that is often unclear to students. In several studies, the use of ICTE in classrooms is characterised as a promising activity to promote science education (Chang, Hsiao, & Chang, 2011; Dori & Belcher, 2005; Scalise, 2012). For example, the work of Webb (2005) presents studies regarding how students use different ICT resources and how these resources contribute to science learning. For this author, the current challenge for teachers and educators is understanding how ICT might potentially support cognitive development, formative assessment and new science curricula. To understand the main ICTE and how they are used to teach science, we organised our analysis into two categories: science teaching and learning through *ICT tools and digital resources* and *the use of digital laboratories* (Table 1.1).

The studies regarding media and hypermedia objects refer to the resources used by teachers to develop science education when the expected result is related to *the learning of science content by students* (Table 1.1). The technologies that emphasise

Table 1.1   Categories and subcategories associated with ICT-based science education

Category	Subcategory
Science teaching and learning through ICT tools and digital resources	Emphasis on simulations and simulation software
	Emphasis on animations
	Emphasis on hypermedia, multimedia, the Web and the Internet
	Emphasis on computer games and serious games
	Emphasis on multi-user virtual environments (MUVEs)
	Emphasis on computer-assisted instruction (CAI)
	Emphasis on specific resources: videos, wikis, podcasts, blogs and chatrooms
	Emphasis on objects: interactive whiteboards (IWB), smartphones and tablets
Science teaching and learning through the use of digital laboratories	Emphasis on virtual laboratories
	Emphasis on remote laboratories and data

digital laboratories are supported by the use of virtual laboratories (along with simulation software) and remote data laboratories, which are current trends in science and technology education.

1.4 How Can Science Teaching and Learning Mediated by ICTE Be Characterised?

1.4.1 Science Education and Learning Mediated by Media and Hypermedia Technologies

Table 1.2 presents emphasis, themes studied and examples that are primarily centred on the effects of ICT on science learning and less on the learning process that occurs (Chang & Tsai, 2005; Lee et al., 2011; Olympiou & Zacharia, 2012).

1.4.1.1 Emphasis on Simulations and Simulation Software

Tables 1.1 and 1.2 show that science teaching and learning are mediated by different digital resources. The first and most cited resource is the use of simulations, which is immediately followed by educational software (Lindgren & Schwartz, 2009; Scalise et al., 2011; Smetana & Bell, 2012). Simulations and educational software were analysed separately from virtual laboratories because not all simulations are used for this purpose. A summary of the main themes investigated regarding this topic is presented in Table 1.2.

Use, Impact and Effects of Simulations on Teaching, Learning and Thinking Skills

According to Lindgren and Schwartz (2009), interactive simulations are powerful tools for scientific thought. Some researchers have studied the effects of simulation on science learning (simulation-based learning). For example, the review by Smetana and Bell (2012) suggests that the use of simulations to teach science may often be more effective than traditional pedagogical practices (based on lectures, books or physical experiments – PE); simulations may assist in the construction of scientific knowledge and the development of skills (e.g. research, data collection or changing variables), in addition to the promotion of an evolution in the students' understanding of scientific concepts.

The review by Scalise et al. (2011) examined the simulation software available for sixth- through twelfth-grade science classes. The review analysed 79 studies and presents a synthesis of the literature, reviews of available products and learning outcomes. The authors determined that 53% of the studies reported overall learning gains, approximately 18% reported gains under certain conditions, approximately 25%

reported mixed results in which some groups exhibited learning gains and other groups did not and 4% reported no gain. An analysis of these results indicates that 96% of the articles in this review presented some type of learning gain when simulations were used and 29% of the studies reported no learning gain in some aspect. These data provide an incentive to understand the real effects of using simulations to teach science because it is currently possible to find free and commercial simulations and virtual laboratory software for use in science teaching (Scalise et al., 2011).

Table 1.2 Main emphasis and themes associated with ICT-based science education

Emphasis	Themes studied	Examples
Emphasis on simulations and simulation software	Use, impact and effects of simulations on teaching, learning and thinking skills	Evagorou, Korfiatis, Nicolaou, and Constantinou (2009), Lindgren and Schwartz (2009), Plass et al. (2012), Reid, Zhang, and Chen (2003), Scalise et al. (2011), Smetana and Bell (2012), Stieff (2011), Veermans, van Joolingen, and de Jong (2006), Zhang, Chen, Sun, and Reid (2004)
	Relationship between theory and reality: the manipulation of variables and modelling	Hansen, Barnett, MaKinster, and Keating (2004), Khan (2010), Li, Law, and Lui (2006), Neulight, Kafai, Kao, Foley, and Galas (2007), Ronen and Eliahu (2000)
	Promotion of scientific argumentation	Chen and Looi (2011), Clark and Sampson (2007)
	Effects of simulation in differentiated classrooms	Dori and Belcher (2005), Rutten, van Joolingen, and van der Veen (2012)
	Simulation as a tool for learning assessment	Quellmalz, Timms, Silberglitt, and Buckley (2012)
	Development and performance of tasks by students using simulations	Olde and de Jong (2004), Ronen and Eliahu (1999)
Emphasis on animations	Effects of animation: motivation to learn science	Barak et al. (2011), Dalacosta et al. (2009), Marbach-Ad, Rotbain, and Stavy (2008), Starbek et al. (2010)
	Development, interpretation and evaluation of animations by students	Chang, Quintana, and Krajcik (2010), Chang, Yeh, and Barufaldi (2010)

(continued)

Table 1.2 (continued)

Emphasis	Themes studied	Examples
Emphasis on hypermedia, multimedia, the Web and the Internet	Internet- and Web-based science learning and their effects: skills, concepts and problem-solving	Chin-Chung (2009), Jang (2006), Lee et al. (2011), She et al. (2012)
	Browsing patterns, reading patterns and webpage search preferences and strategies	Dimopoulos and Asimakopoulos (2009), Gelbart et al. (2009), Hoffman, Wu, Krajcik, and Soloway (2003), Ng and Gunstone (2002), Shapiro (1999), Zhang (2012)
	Researching and sharing data online	Kubasko, Jones, Tretter, and Andre (2008), Mistler-Jackson and Butler Songer (2000)
	Effects of a hypermedia system on science learning	Jacobson and Archodidou (2000), Liu and Hmelo-Silver (2009), Moss and Crowley (2011), Schaal, Bogner, and Girwidz (2010)
	Effects of incorporating a multimedia system into science classes: the relationship between attention and eye movement, actions required by the teacher, student's critical thinking, representations of scientific phenomena, problem-solving, scientific reasoning, conceptual change, motivation and its relationships and scientific investigation	Ardac and Akaygun (2004), She and Chen (2009), She and Lee (2008), She and Liao (2010), Shin, Jonassen, and McGee (2003), Tseng, Tuan, and Chin (2010), Waight and Abd-El-Khalick (2007), Zheng et al. (2008), Zydney and Grincewicz (2011)
	Student-authored multimedia and steps for the development of a constructivist multimedia	Orion, Dubowski, and Dodick (2000), Tekos and Solomonidou (2009)

(continued)

Table 1.2   (continued)

Emphasis	Themes studied	Examples
Emphasis on computer-assisted instruction (CAI)	Student behaviour: responsibility and the relationship between the genders	Mayer-Smith, Pedretti, and Woodrow (2000), Pedretti, Mayer-Smith, and Woodrow (1998)
	Improvements in performance, attitudes and career aspirations in relation to science	Hansson, Redfors, and Rosberg (2011), Park, Khan, and Petrina (2009)
	Self-regulated learning process	Devolder, van Braak, and Tondeur (2012)
	Reduction in misconceptions and partial explanations	Hsu, Wu, and Hwang (2008)
	Collaborative actions: analysis, synthesis and testing interpretations	Ebenezer, Kaya, and Ebenezer (2011), Ergazaki, Zogza, and Komis (2007)
	Conceptual and theoretical understanding	Barak and Dori (2005)
	Scientific argumentation, modelling and collaborative work	Oshima et al. (2004)
	Types of CAI: student- and teacher-centred	Chang and Tsai (2005), Pol, Harskamp, and Suhre (2005)
	Tutorial systems and problem-solving: students' capacity for problem-solving, exploration and planning and comparison with science textbooks	Pol et al. (2005), Soong and Mercer (2011)
Emphasis on computer games and serious games	Prediction-observation-explanation strategy	Hsu (2006), Hsu, Tsai, and Liang (2011)
	Knowledge gains and impact on learning	Klisch, Miller, Wang, and Epstein (2012), Squire and Jan (2007)
	Changing attitudes	Klisch et al. (2012)
	Scientific thinking and argumentation	Squire and Jan (2007)

(continued)

Table 1.2 (continued)

Emphasis	Themes studied	Examples
Emphasis on multi-user virtual environments (MUVEs)	Reflections and narrative on conceptual, socio-science and science ethics issue	Barab, Sadler, Heiselt, Hickey, and Zuiker (2007), Barab et al. (2009), Cher Ping (2008), Furberg and Ludvigsen (2008)
	Active engagement in learning processes and academic motivation	Barab et al. (2007, 2009), Cher Ping (2008), Hakkarainen (2003), Lim, Nonis, and Hedberg (2006)
	Complex system thinking skills	Evagorou et al. (2009)
	Guidelines for the use of a MUVE and its influence on learning	Nelson (2007)
	Reflection on curricular organisation	Barab et al. (2009)
	Scientific research process and collaborative learning	Hakkarainen (2003); Lin, Hsu, and Yeh (2012); Tolentino et al. (2009)
	Student perceptions regarding participation in a MUVE	Rosenbaum, Klopfer, and Perry (2007)
Emphasis on specific resources: videos, wikis, podcasts, blogs and chatrooms	Podcasts as an ubiquitous means of education	Holbrook and Dupont (2011)
	Photos as a tool to explore ideas and promote scientific argumentation	Byrne and Grace (2010), Katz (2011), Piburn et al. (2005)
	Use of videos to facilitate questioning, the creation of meaning and promoting changing ideas	Furman and Barton (2006), Mayo, Sharma, and Muller (2009)
	Chatrooms as a means to exchange information for science learning	Pata and Sarapuu (2006)
Emphasis on interactive whiteboards (IWB), smartphones and tablets	Tools for mobile learning and data sharing	Looi et al. (2011), Zhang et al. (2010)
	Interactive whiteboard (IWB) as a shared dialogic space	Warwick et al. (2010)
Emphasis on virtual laboratories	Combination of virtual resources with real experiments	Clark and Jorde (2004), Jaakkola and Nurmi (2008), Jaakkola, Nurmi, and Veermans (2011), Olympiou and Zacharia (2012), Zacharia, Olympiou, and Papaevripidou (2008)
	Implicit and explicit instruction in real and virtual laboratories	Jaakkola et al. (2011)
	Pedagogical proposals for the integration of virtual laboratories: case studies and constructivist teaching strategies	Dori and Sasson (2008) and Russell et al. (2004)
	Preferences for the use of virtual laboratories	Sun, Lin, and Yu (2008)

(continued)

Table 1.2 (continued)

Emphasis	Themes studied	Examples
Emphasis on remote laboratories and data	Effects of remotely controlled laboratories	Kong, Yeung, and Wu (2009), Lowe, Newcombe, and Stumpers (2012), Underwood, Smith, Luckin, and Fitzpatrick (2008)

Some researchers have taken a particular interest in the effects of simulation software on learning concepts and understanding scientific phenomena (Dori & Belcher, 2005; Lindgren & Schwartz, 2009; Stieff, 2011). Detailed studies on student perceptions and spatial learning through science simulations are presented in the work of Lindgren and Schwartz (2009).

Veermans et al. (2006) investigated the use of two learning environments in the physics field of collisions (one environment with implicit heuristics and the other environment with explicit heuristics). The question of the heuristic's presence is related to the guidance required for students to develop simulation-based activities. The results of Veermans et al. (2006) suggest that the explicit presentation of the heuristics facilitates more self-regulation in the students. In addition to the presence of heuristics, another suggestion is that the simulations should be based on a *discovery learning* proposal, which is organised in a threefold scheme: interpretative support (IS), experimental support (ES) and reflective support (RS). According to Reid et al. (2003) and Zhang et al. (2004), learning support in a simulation environment (SE) should be directed at these three perspectives for discovery learning to be meaningful, systematic and reflective.

Relationship Between Theory and Reality: The Manipulation of Variables and Modelling

The use of simulations to relate theoretical studies to the real world also appears in several works (Khan, 2011; Neulight et al., 2007; Ronen & Eliahu, 2000). According to Khan (2011), computer simulations can be particularly attractive for science students because they can observe changes in phenomena via the manipulation of variables. For example, to assess the potential of an electric circuit simulation that helps students relate theory to reality, Ronen and Eliahu (2000) demonstrated that this type of simulation helps students identify and correct their mistakes and address common problems of relating formal representations to real circuits and vice versa. The same result was observed in the work of Neulight et al. (2007), who investigated student understanding of natural infectious diseases based on an understanding of a virtual infectious disease.

Other examples of studies that relate theory to reality are studies related to modelling software (Hansen et al., 2004; Li et al., 2006). The increased availability of computer modelling software has created opportunities for students to participate in scientific investigations via the construction and analysis of computer-based models of scientific phenomena. However, despite the growing trend of integrating technology into science curricula, it is important for educators to understand the specific aspects of these technologies that promote student learning. According to Hansen et al. (2004), knowledge evolves when it is actively constructed compared with the passive acceptance of theories and relationships that have previously been tested.

Promotion of Scientific Argumentation

Educational software can also be used to promote scientific argumentation (e.g. Clark and Sampson (2007) investigated personally seeded discussions (PSD) to scaffold online argumentation). The study of Chen and Looi (2011) sought to examine the nature of teacher-student and student-student discourse when leveraged by a software called Group Scribbles (GS), which was used to study the dispersion of fruits and seeds in a fifth-grade science classroom. GS promoted group learning and helped students improve the quality of their ideas during the construction and reconstruction of scientific knowledge.

Effects of Simulation in Differentiated Classrooms

The study of Dori and Belcher (2005) analyses the effects of a media-rich, differentiated learning space, referred to as the Technology-Enabled Active Learning (TEAL) project, on the cognitive and affective outcomes of students at the Massachusetts Institute of Technology (MIT). The project involved software for simulation and visualisation of physical phenomena and processes, which were conducted in a classroom that had been redesigned to facilitate group interactions. The research population comprised 811 undergraduate students divided into small- ($n = 176$) and large-scale ($n = 514$) experimental groups and a control group ($n = 121$). The project's assessment included an examination of the students' conceptual understanding before and after studying electromagnetism in this space and also investigated the project's effect on the students' preferences for different teaching methods. The results indicated that the students who studied in the TEAL format significantly improved their conceptual understanding of different complex phenomena associated with electromagnetism. Most students in the small-scale experiment would recommend the TEAL course to their peers, which indicates the benefits of interactivity and the visualisation of virtual practical experiments. In the large-scale implementation, the students expressed positive and negative attitudes towards the course. The positive attitudes were related to knowledge gains, and the negative attitudes were related to acceptance of the project; in these types of basic Physics courses, students have traditionally been accustomed to primarily lecture-based classes, having passive attitudes, closely following a textbook and studying for exams.

Science is typically taught within a traditional school setting. Not all schools offer active and innovative spaces, such as in the work of Dori and Belcher (2005). Thus, how can computer simulations be used in this teaching context to improve learning processes and outcomes? This question is present in a review by Rutten et al. (2012), in which they compare teaching conditions with or without simulations. The study indicates positive results where simulations were used to replace or enhance traditional teaching, especially regarding the performance of activities that precede a physical laboratory (pre-laboratory activities). Similar results are indicated in the study of Plass et al. (2012), which demonstrated the use of simulations in a sequence based on the conceptual development of ideas improved the students' chemistry performance. In light of these results, we can infer that the integration of simulations in traditional science classes has a positive trend for student learning.

Simulation as a Tool for the Assessment of Learning

The use of educational software in science teaching can also be used to assess the students rather than to teach them. The work of Quellmalz et al. (2012) presents a system for the assessment of simulation-based science learning: SimScientists simulation. This proposal was tested with 5687 secondary school students and 55 science teachers. The results demonstrated that the students performed better on assessments that were based on an interactive simulation compared with the conventionally proposed static content items (ecosystem, force and motion) at the post-test. However, more research is needed to document the effectiveness of this type of assessment to support learning, especially about its planning and use by science teachers.

Development and Performance of Tasks by Students Using Simulations

Continuing on the subject of assessment, we can contemplate the possibility of students also *developing* simulation-based tasks. Olde and de Jong (2004) investigated the development of tasks using simulations by 19 first-year technical education students as a knowledge-generating activity. The results indicated that the students not only designed the assignments according to facts or procedures but also observations made with the simulation. During the development process, the students reinforced their prior knowledge, recovered and explained the steps for solving problems and focused on the dynamic characteristics of the simulated circuits.

In general, we realised that the use of simulations for science teaching and learning has important components: interactive work (Dori & Belcher, 2005; Khan, 2011), cooperative work (Hansen et al., 2004) and feedback (Chen & Looi, 2011). Not all studies identify positive results for the use of simulations to teach science. For example, the study by Stieff (2011), which sought to identify the effect of a Connected Chemistry Curriculum (CCC), analysed the content knowledge of 460 secondary students taught by four chemistry teachers, as well as their representational competence in chemistry, using simulations incorporated in investigative activities. The results indicated that the simulations that accompanied the CCC produced only modest gains in student performance when summative evaluations were conducted. Furthermore, the students were significantly more likely to use representations consistent with the representations used by the teacher and specialist.

However, in most of the studies presented, the use of simulations has been addressed without taking into account the potential impact of teacher support, the classroom setting and the place of simulations within the curriculum (Rutten et al., 2012; Scalise et al., 2011; Smetana & Bell, 2012).

1.4.1.2   Emphasis on Animations in an Isolated System

For the second group of resources in this chapter, the rapid growth of screen use (e.g. tablets, personal computers and smartphones) has also influenced science education. A number of science researchers and educators believe that the use of *animations* that represent scientific models has great potential to support the teaching and learning of scientific concepts (Barak et al., 2011; Marbach-Ad et al., 2008; Starbek et al., 2010). For Marbach-Ad et al. (2008) animation and simulation are the same; for us, they are distinct concepts and resources, despite very similar properties. We rely on the concept of Höffler and Leutner (2007), wherein "an animation can be defined as a series of rapidly changing computer screen displays suggesting movement to the viewer" (p. 723). Table 1.2 presents a summary of the main results of the studies that investigated the use of animations to teach science.

Effects of Animation: Motivation to Learn Science

The studies of Barak et al. (2011) and Dalacosta et al. (2009) investigated the effects of animated flash movies on student learning outcomes. For Barak et al. (2011), the use of cartoons provided the students with a greater motivation to learn science, a capacity for explanation and an understanding of scientific concepts compared with the control group. Dalacosta et al. (2009) also demonstrated that an understanding of specific scientific concepts using animation is more effective in elementary education students and can be used as an additional didactic tool for teachers at this education level.

Starbek et al. (2010) investigated whether the use of animations to *teach genetics* contributed more to the students' knowledge and understanding than other modes of teaching. To this end, the researchers conducted one pre-test and two post-tests. A quasi-experimental study was conducted with four comparable groups: group 1 was taught in the traditional classroom lecture format ($n = 112$), group 2 only read texts ($n = 124$), group 3 watched two short computer animations ($n = 115$) and group 4 read a text with illustrations ($n = 117$). The authors determined that the best learning outcomes were obtained through the use of animations (group 3) followed by illustrated texts (group 4). Thus, this study demonstrates that traditional classroom lectures, followed by reading texts, have a low contribution to the construction of scientific knowledge compared with the use of animated resources. The study of Marbach-Ad et al. (2008) complements the study of Starbek et al. (2010). These researchers recommended the use of animations, particularly when teaching dynamic scientific processes, because their use can improve student performance in comparison with traditional teaching.

Development, Interpretation and Evaluation of Animations by Students

Student involvement in activities with cartoons was also examined in the studies of Chang, Yeh, and Barufaldi (2010) and Chang, Quintana, and Krajcik (2010), which investigated the impact that the design and evaluation of molecular animations had on how 271 seventh-grade students understood the particle nature of matter. The students were randomly divided into three groups that used *Chemation*, a student-centred animation tool to (1) design, interpret and evaluate animations, (2) only design and interpret animations or (3) only view and interpret animations made by teachers. The results indicate that designing peer-evaluated animations is effective in improving student learning. However, the effectiveness of allowing students to design animations without peer evaluation is questionable compared only with the viewing of animations by students. This study did not investigate other combinations of animation-based modelling activities or the teacher's role in supporting student learning with technology.

1.4.1.3 Emphasis on the Internet, Web, Hypermedia and Multimedia

Educators and researchers have argued about the effectiveness of hypermedia, multimedia and the World Wide Web (WWW) to facilitate viewing different science content (Mistler-Jackson & Butler Songer, 2000; She & Liao, 2010; Tseng et al., 2010; Zheng et al., 2008). The main emphasis is shown in Table 1.2.

Internet- and Web-Based Science Learning and Their Effects: Skills, Concepts and Problem-Solving

"Internet-based science learning" has been championed by many educators for more than a decade. Lee et al. (2011) conducted an important review of journal articles on *Internet-based science learning* from 1995 to 2008. Several important conclusions are drawn. For example, letting the students take control is essential to improve their attitudes and motivation for Internet-based science learning. However, proper guidance from teachers, moderators or Internet-based learning environments remains very crucial in science learning via the Internet.

It is important to note that the Internet has become a conduit for other forms of scientific content dissemination in a hypermedia format. Thus, considering the evidence, would integrating the Web into science classes improve student learning? She et al. (2012) investigated the effects of solving chemistry problems using the Web, and Jang (2006) investigated the effects of incorporating the Web into science classes on four groups of seventh-grade students. The latter work sought to understand student performance and attitudes towards the use of the traditional teaching method compared with Web-based teaching to study *heat, light reflection* and *life changes*, among other science content. Through mixed methods and the quasi-experimental method, the average result of the students' final exams was higher with the experimental teaching method compared with the students who received traditional teaching. The two teaching methods exhibited significant differences with regard to student achievement. Once again, we must be cautious with these results because the study had limitations in its data collection. Even considering these limitations, more than half of the students preferred the experimental method compared with traditional teaching.

Browsing Patterns, Reading Patterns and Webpage Search Preferences and Strategies

Another important topic refers to webpage search strategies. A study by Dimopoulos and Asimakopoulos (2010) explored the webpage browsing preference patterns of ten secondary school students conducting online research regarding "Cloning". The results demonstrated that most students browse the Web in a very superficial way. They tend to visit a large number of pages (approximately 20 pages in less than an hour) but remain on only a few pages (approximately three or four pages) for an extended period of time, which allows them to analyse the content in a more concentrated way. Zhang (2012) notes that the online reading of scientific content is typically superficial and fragmented and needs to be guided.

For Shapiro (1999), students create hierarchical representations as they work on webpages, which are important to define relationships between the concepts. Although the Web has a number of positive effects on student learning, including motivation for independent learning (Ng & Gunstone, 2002) and metacognitive search strategies (She et al., 2012), its unedited and unstructured nature suggests that many sites that students visit have information that is also difficult to understand. For students to research effectively, improve their technical skills and critical thinking and learn scientific content on the Web, they need time to research, appropriate strategies, carefully chosen resources and either teacher guidance or a structured system of guidelines (Gelbart et al., 2009; Hoffman et al., 2003; Ng & Gunstone, 2002; Shapiro, 1999; Zhang, 2012).

Researching and Sharing Data Online

We also identified a number of science projects for sharing data online (Hoffman et al., 2003; Kubasko et al., 2008; Mistler-Jackson & Butler Songer, 2000). In these projects, a group of individuals, such as students, teachers or scientists, share data and collaborate on scientific questions and current events. Data sharing programs that use network technologies differ in many aspects, including the target audience, their objective, the use of technological tools, teacher autonomy and the flexibility of data collection standards. Mistler-Jackson and Butler Songer (2000) studied how different sixth-grade students view content for learning about the *atmosphere* and the use of technology before and after their participation in the "Kids as Global Scientists (KGS)" program, which used data available on the Internet. The results indicated that the students achieved significant gains in knowledge related to "Weather" when what they wrote was assessed, and the interviews indicated a high level of motivation and satisfaction related to participation in the project. These common characteristics also include communication, collaboration, authenticity and access to real-time information.

Effects of a Hypermedia System on Science Learning

We also identified studies on the use of hypermedia for science training, teaching and learning (Jacobson & Archodidou, 2000). In particular, the work of Liu and Hmelo-Silver (2009) presents results on how a *conceptual representation*, through a hypermedia system, influenced the training of 82 teachers and developed a deeper understanding of the "human respiratory system" in 41 seventh-grade students. Two versions of educational hypermedia were created: one version focused on the respiratory system's function and behaviour, and the other version focused on this system's structure. The results of both studies demonstrated that the participants who used the function-centred hypermedia (F-hypermedia) developed a deeper understanding compared with the participants who used the structure-centred version (S-hypermedia). This study suggests that a conceptual representation centred on the function of scientific systems is a method to promote an understanding of complex systems in science teaching that deserves greater attention.

Effects of Incorporating a Multimedia System in Science Classes

Multimedia environments enable the association of different resources (Zydney & Grincewicz, 2011). In the context of using multimedia systems to teach science, research with different objectives and results was identified. For example, the study of She and Chen (2009) examined how secondary students constructed their understanding of *mitosis* and *meiosis* at a molecular level through multimedia educational materials presented in different interactions and sensory modes (animation/narrated simulation/on-screen text). The study's results indicated that the group that received simulation with on-screen text allocated a greater amount of visual attention than the group that received animation with on-screen text and narrated simulation. For She and Chen (2009), eye movement is closely related to science learning. Thus, this study adds empirical evidence of a direct correlation between the duration of a fixed gaze and learning depth.

Ardac and Akaygun (2004) and Zheng et al. (2008) suggest that the effectiveness of a multimedia-based environment can be improved when this resource is integrated with pedagogical planning for teaching and if the instruction includes an additional suggestion to students with different representations of the same phenomena, for example, scientific argumentation, collaborative work and research followed by reflection. Its effective or ineffective use will often also be associated with the teacher's actions; they are not always prepared or may not have had appropriate training for the use of ICT in the classroom (Barak & Dori, 2011; Kumar et al., 2011; Yarden & Yarden, 2011). Researchers suggest that the use of multimedia resources to teach science contributes to scientific reasoning, the motivation to learn science and students' conceptual change (She & Lee, 2008; She & Liao, 2010; Shin et al., 2003; Tseng et al., 2010). However, from the data collected, we note that an *excess* of multimedia resources may constrain the students' learning process. For example, a study by Waight and Abd-El-Khalick (2007) assessed the impact of a multimedia environment on the representation of scientific "investigation" in 42 sixth-grade students. The results indicated that the multimedia resource used (microcomputer-based laboratories; simulations and microworlds; telecommunication technologies, including e-mail and Internet interfacing, as well as accessing and using Web-based databases) worked to restrict rather than promote "inquiry" during classroom participation. In the presence of computers, group activities became more structured and focused on task sharing, and less time was devoted to group discourse.

Student-Authored Multimedia and Steps for the Development of a Constructivist Multimedia

With respect to the potential of a multimedia authoring program as a science learning tool, the research of Orion et al. (2000) demonstrated that although most students liked to use the multimedia program, there was no evidence to support the hypothesis that it contributed to the students' acquisition of scientific knowledge (32 students in two twelfth-grade classes). In fact, most of the time they spent authoring multimedia was dedicated to the production of decorative effects, which reduced the time available for meaningful learning. As a solution, the authors suggest that an integration of laboratory exercises, visits and an independent study project could lead to meaningful learning; however, these actions also depend on how the teacher conducts the activities.

The work of Tekos and Solomonidou (2009) provides significant guidance for the development of a "Constructivist Multimedia Learning Software" following the ICDDI (investigation, conception, design, development and implementation) model, which describes the steps that should be followed to create, implement and evaluate constructivist learning environments. The striking difference between the work of Orion et al. (2000) and Tekos and Solomonidou (2009) is in the structured nature of the activity and how the teacher conducts it. These actions are important to the process of learning science through a multimedia system.

1.4.1.4   Emphasis on Computer-Assisted Instruction (CAI)

Several studies have characterised computer-mediated science education or CAI. This is the integration of ICT in science teaching, with an emphasis on computers (Ergazaki et al., 2007; Hsu et al., 2008). Different studies aim to investigate the potential of CAI for teaching science and are characterised in Table 1.2.

Main Effects of CAI on Teaching Science

One of the first studies on CAI is the work of Pedretti et al. (1998), which analysed teaching, learning and this technology's impact from the perspective of 144 secondary school students who participated in a study to implement CAI. The most surprising data from this study were that the students exhibited greater attention to issues related to learning rather than technology or science. This same group of researchers (Mayer-Smith et al., 2000) also investigated that the pedagogical practices and social organisation in science classes mediated by technology can promote an inclusive experience of gender, in which boys and girls participate and perform the proposed activities equally well.

The work of Devolder et al. (2012) provides a literature review that covers different supports for self-regulated learning (SRL) processes in the field of computer-mediated science education. The most effective supports are categorised and discussed according to the different areas and stages of SRL: student cognition, motivation, behaviour and context. The results indicate that most studies on "scaffolding" processes focus on cognition, whereas few studies focus on the noncognitive areas of SRL.

Other effects of CAI are also demonstrated in the work of Barak and Dori (2005), which investigated the impact of IT-assisted project-based learning (PBL) on the achievements of chemistry students and their ability to navigate the different understanding levels of chemical concepts (symbolic, macroscopic, microscopic and procedural). The chemistry students who participated in the IT-mediated PBL ($n = 95$) had a significantly better performance than the control group ($n = 120$), not only at post-test but also on the final course exam. More generally, the results indicated that the incorporation of PBL within an IT-rich environment may improve students' understanding as they study chemical concepts, theories and molecular structures.

Finally, an important study is the work of Oshima et al. (2004), which characterises student behaviour in a CAI environment. The research of Oshima et al. (2004) reported the development of two scientific experiments, "how things burn" and "how a candle goes out in a closed flask", in a sixth-grade classroom using Computer-Supported Collaborative Learning (CSCL). In both experiments, the CSCL technology was primarily implemented to facilitate group collaborations. The results indicated the following: (1) the students were more likely to engage in symmetrical communication (i.e. among the groups as well as within the groups) in the second experiment, and (2) they were also more focused on ideas and more often shared their ideas in the second experiment. The results were discussed from the perspectives of the scientific practices of the students engaged in the task structure.

Types of CAI: Student- and Teacher-Centred

The research of Oshima et al. (2004) demonstrated that CAI was student-centred. Would the results be different if CAI was teacher-centred? A study by Chang and Tsai (2005) investigated the effects of two types of CAI, teacher-centred (TCCAI) and student-centred (SCCAI), on student learning outcomes and their personal learning environment (PLE). Three hundred forty-seven tenth-grade students participated in the study. One group of students ($n = 216$) was taught by TCCAI, whereas the other group of students ($n = 131$) was taught by SCCAI. The results indicated that (a) there were no significant differences in the performance of student activities between the two groups, (b) the TCCAI group had significantly better attitudes towards science content than the SCCAI group and (c) a significant PLE-treatment interaction was identified for student attitudes towards the subject matter; the teacher-centred instructional approach appeared to enhance more positive attitudes of less constructivist-oriented learning preference students, whereas the student-centred method was more beneficial to more constructivist-oriented learning preference students on their attitudes towards earth science in a computer-assisted learning environment (Chang & Tsai, 2005).

Tutorial System and Problem-Solving

Chang and Tsai (2005) found no effects from CAI (teacher-centred versus student-centred) compared with the students' use of science textbooks. Pol et al. (2005) observed the use of hints and the students' ability to identify a solution to a physics problem both with the help of CAI, in this case, a program called NatHint, and with the help of a science textbook. Thirty-six secondary school physics students participated in this study: an experimental group ($n = 11$) that used their textbook and the computer program and a control group ($n = 25$) that used only the book. The results indicated there was no evidence that the students in the experimental group achieved better results for problem-solving. The students' exploration and planning ability was improved. The students involved in the experiment made better use of their declarative knowledge in problem-solving compared with the students in the control group.

1.4.1.5   Emphasis on Computer Games and Serious Games

Over the decades, the impact of computer games has been analysed according to various dimensions, which have mainly focused on entertainment, training and learning. The work of Connolly, Boyle, MacArthur, Hainey, and Boyle (2012) presents a literature review of computer and serious games concerning the positive impacts of gaming on users 14 years of age or older, particularly with regard to learning, skill enhancement and engagement. One hundred twenty-nine studies were identified, and "the findings revealed that playing computer games is linked to a range of perceptual, cognitive, behavioral, affective and motivational impacts and outcomes" (Connolly et al., 2012). Thus, we were interested in examining how computer games and serious games contribute to science teaching and learning. In response to this question, Table 1.2 presents a summary of the main effects identified in the studies on the use of computer games to teach science.

Main Effects of Computer Games on Science Classrooms

Hsu (2006) and Hsu et al. (2011) investigated the effects of implementing a computer game that integrated the prediction-observation-explanation (POE) strategy to facilitate preschool children's acquisition of scientific concepts related to *light and shadows*. The study indicated that the students in the experimental group significantly outperformed their counterparts in terms of their understanding of "how shadows are made in daylight" and the "orientation of shadows". However, after playing the games, the children in both groups still expressed some conceptions, for example, "shadows always appear behind a person" and "shadows should be on the same side of the sun". One limitation of the study was that the researchers did not conduct a pre-test to understand the children's conceptual conflicts or misconceptions. This control condition difficulty also appears in the work of Klisch et al. (2012), which investigated knowledge gains and attitude changes attributable to an online science education game called Uncommon Scents. After the game, the students exhibited significant science content knowledge and an attitude change towards inhalants (toxic chemicals).

There are also studies that relate science education with augmented reality games (Enyedy, Danish, Delacruz, & Kumar, 2012; Squire & Jan, 2007). For example, Squire and Jan (2007) investigated whether augmented reality games on portable devices can be used to engage students in scientific thinking (particularly in argumentation), how the game's structure affects the students' thinking and the impact of the game's role on learning, as well as the physical environment's role in shaping learning. The results demonstrated that these games have the potential to engage students in a meaningful scientific argumentation of scientific phenomena. Regarding the research of Squire and Jan (2007), we must be careful with the results because the research activity proposed by the game had a short duration, which was substantially smaller and shorter than real research; there was a lack of systematic data from a pre-/post-test concerning the students' performance; and the investigators and monitors had an active role in supervising the game. The limitations of Squire and Jan (2007) are valid for several studies that use computer games in science education.

1.4.1.6 Emphasis on Multi-user Virtual Environments (MUVEs)

One interactive medium that has been the focus of recent research and has complemented computer games is educational MUVEs. Educational MUVEs have emerged in recent years as a promising platform for science learning. Table 1.2 shows the main themes presented in studies regarding the use of MUVEs to teach science (Barab & Dede, 2007; Barab et al., 2009; Barab, Sadler, Heiselt, Hickey, & Zuiker, 2007; Dede & Barab, 2009; Ketelhut, 2007).

With regard to the first study, Barab et al. (2007) investigated the potential of the MUVE educational game Quest Atlantis (see also Cher Ping, 2008; Lim et al., 2006) in relation to producing a socio-science narrative with 28 fourth-grade students and this game's interactive role to support learning. The MUVE tells the history of a city facing ecological, social and cultural decay (similar to existing global issues) because of its rulers' blind pursuit of prosperity and modernisation. In general, the students were engaged with the activity, promoted rich scientific discourse, presented quality work and learned the science content. Furthermore, through their participation with the narrative, the students developed a rich perceptual, conceptual and ethical understanding of science. The study of Furberg and Ludvigsen (2008) completes this idea, that is, for students to gain a deeper understanding of socio-scientific issues in environments mediated by ICT.

Few studies have investigated the use of MUVEs and student performance in different science teaching models (Evagorou et al., 2009). We can cite the study of Barab et al. (2009) as an example in which the researchers studied society's use of agricultural pesticides and waste production through a MUVE with 51 undergraduate students. In this study, the researchers compared pairs of students randomly distributed to four different teaching conditions, in which the content became increasingly more contextualised: (a) an expository condition with an electronic textbook, (b) a simplistic framework condition, (c) an immersive world condition with two users and (d) an immersive world condition with a single user. The results indicate that the two users (c) and single user (d) in an immersive world conditions had a significantly better performance than the group with the electronic textbook and the simplistic framework condition. The two users in an immersive world condition also had a significantly better performance than the other teaching conditions. According to the researchers, MUVEs provide a powerful new way to develop a science curriculum.

Another scenario we may present refers to MUVEs' ability to promote research on scientific phenomena (Hakkarainen, 2003; Lin et al., 2012). The study's results indicate that with the teacher's guidance, the students were able to produce significant intuitive explanations regarding biological phenomena, participate in this process, pursue their own research questions and engage in a constructive peer interaction that helped them move beyond their intuitive explanations towards scientific explanations.

1.4.1.7   Emphasis on Specific Resources: Wikis, Podcasts, Blogs, Photos and Videos

We identified several works on the development of wikis (Kim & Herbert, 2012), podcasts (De Winter, Winterbottom, & Wilson, 2010; Holbrook & Dupont, 2011) and blogs (Wang, Ke, Wu, & Hsu, 2012) as resources that promote science teaching and learning. We summarise the main themes in this subcategory in Table 1.2.

Podcasts as an Ubiquitous Means of Education

The use of podcasts appears in the study of Holbrook and Dupont (2011), which compared the response of 350 first-year undergraduate biology students with 300 second-year students when asked about the use of podcasts available after each *genetics* class. Both the first year and more advanced students reported that the podcasts were very useful for a wide range of learning activities; however, more than 50% of the students reported that the podcasts' availability influenced their decision to skip class. The decision to miss class was more influential for the first-year students.

Photos as a Tool to Explore Ideas and Promote Scientific Argumentation

Byrne and Grace (2010) used a concept mapping tool with a photo association technique to extract ideas from 169 secondary school students (11 years old) about microbial activity. According to the researchers, this tool can be used to explore ideas and encourage scientific argumentation on this and other scientific concepts for children, as well as other age groups; however, language can be a barrier for the participation of some students. Katz (2011) studied how "Photobooks", using the Internet, can improve scientific identity in "early childhood", and Piburn et al. (2005) studied the role of viewing computer-based images for science learning. The results of these studies demonstrate that students' spatial ability and content learning can be improved through instruction, in addition to eliminating performance differences between genders. These results demonstrate that the use of an image followed by instruction has a better effect than the use an image alone.

Use of Videos to Facilitate Questioning, Creation of Meanings and Promotion of Changing Ideas

About the use of videos in science classrooms, we have some interesting results. The work of Mayo et al. (2009), which was conducted with 29 first-year physics students, explores the use of small video segments interspersed with discussions in two settings: a whole class and small discussion groups. The results indicate that the use of video segments in both interventions was successful in changing the students' understanding of *superconductivity*. However, small groups tend to facilitate questions, develop meanings and promote changing ideas more than discussions with the entire room. The study of Furman and Barton (2006) sought to understand the roles that the voices of two seventh-grade students can play when developing a science mini-documentary. This study demonstrates that the integration of the student's voice in a science project can be a valuable tool to shape the students' identity.

Chatrooms as a Means to Exchange Information for Science Learning

The use of virtual chatrooms for "Models and Modeling to Support Science Learning" appears in the work of Pata and Sarapuu (2006). The authors compared the reasoning processes of 53 secondary school students in a "Collaborative Modeling Environment" to learn about genetics problems using virtual chatrooms. In the work of Pata and Sarapuu (2006), the students' problem representation level was measured according to their initial attempts at problem-solving, which resulted in three subgroups of analysis: concrete, semiabstract and concrete-abstract. The activities were supported by a tutor using a virtual chatroom, which promoted student involvement and participation. Some resources, such as photos, videos, wikis and chatrooms, are also currently used to develop "inquiry-based instruction". The use of chatrooms also supports science teaching by communicating results directly via scientific argumentation or for the development of a collaborative research group.

1.4.1.8 Emphasis on Objects: Interactive Whiteboards (IWB), Smartphones and Tablets

We placed studies regarding the use of certain objects in science classrooms into a single subcategory. The objects included IWB (Warwick et al., 2010) and mobile technologies, such as tablets and smartphones (Looi et al., 2011; Zhang et al., 2010). The results of the studies indicate that the teacher's presence is crucial to good science learning outcome, and the research process focuses more on resources and their effects and less on the process. We summarise the results in Table 1.2.

Tools for Mobile Learning and Data Sharing

Research regarding the use of smartphones and tablets is related to the topic "Mobile Technologies" or "Mobile Learning" (Looi et al., 2011). For example, Looi et al. (2011) sought to test and refine a research project using mobile technologies in the third-grade science curriculum of a primary school to study the *human body*. The results indicated that the experimental class performed better than the other classes, which used traditional teaching and assessments. In the classes supported by the use of mobile technologies, the researchers determined that the students learned science in a more personal and involving way, with positive attitudes towards science teaching. The research of Zhang et al. (2010) analysed the effects of mobile technologies when 39 primary school students used them routinely in science education. There was a positive change in the teacher's teaching practice and the students' attitudes towards the use of smartphones. The students were engaged in research tasks, such as data collection and group work. Smartphones are still underutilised as classroom resources because not all teachers know how to integrate them in their lessons. However, tablets have already gained ground because they are used as digital books and resources for note taking and storing images, as well as for Internet access, animations and simulations of scientific phenomena.

IWB as a Shared Dialogic Space

Warwick et al. (2010) investigated how students use an IWB when working on scientific activities. An IWB has the potential to encourage the creation of a shared dialogic space, within which collaborative knowledge construction can occur. However, this only occurs when there is active support from the teacher for the collaborative and dialogical activity in the classroom, in which the teacher is able to devise tasks that take advantage of opportunities to promote active student learning (Warwick et al., 2010).

1.4.2   Science Teaching and Learning Mediated by Virtual and Remote Laboratories

Given the limited number of laboratories and experimental science devices, primary school students in many countries typically do not have sufficient opportunities to conduct scientific experiments during class time (Dori & Sasson, 2008;Kong et al., 2009 ; Lowe et al., 2012). In response to this problem, many investigations have emerged regarding the potential to integrate ICT to enable students to conduct scientific experiments (Underwood et al., 2008). We were able to identify two promising developments in this aspect: the first development refers to virtual laboratories, and the second development refers to remotely controlled laboratories or remote laboratories. We summarise the main results of the research regarding the use of virtual and remote laboratories in Table 1.2.

1.4.2.1   Emphasis on Virtual Laboratories

Over the previous few decades, different studies have investigated the use of physical experiments (PEs) and virtual experiments (VEs) in scientific experimentation laboratories (Jaakkola et al., 2011; Jaakkola & Nurmi, 2008; Pyatt & Sims, 2012;

Russell, Lucas, & McRobbie, 2004; Zacharia, 2007; Zacharia et al., 2008). We identified several examples regarding the use of virtual laboratories to teach science, which deserve to be discussed.

Combination of Virtual Resources with Real Experiments

Comparative studies were performed to identify which of the two modes of experimentation (PE or VE) is the most widely used in different scientific fields. Many science teachers ask whether it is better to combine simulation activities with laboratory activities or use the resources separately. Jaakkola and Nurmi (2008) investigated the best combination to teach *simple electricity* concepts to fifth-grade students. The results indicated that the combination of virtual simulation with laboratory resources led to more significant learning gains than the use of any simulation or laboratory activity in isolation, and it also more efficiently promoted the students' conceptual understanding. There were no significant differences between the simulation and laboratory environments.

Zacharia et al. (2008) confirmed the results of Jaakkola and Nurmi (2008). These authors investigated the comparative value of experiments with physical manipulatives (PM) in a sequential combination with virtual manipulatives (VM), with the use of PM preceding the use of VM, and experiments with PM in isolation regarding changes in 62 undergraduate physics students' conceptual understanding of "heat and temperature". The results indicated that experimentation with a combination of PM and VM improves the students' conceptual understanding more than experimentation with PM alone. The use of VM was identified as the cause of differentiation; however, not all studies exhibit the same results with the use of a combination of real experiments and VE. For example, in the study of Pyatt and Sims (2012), chemistry undergraduate students who conducted an investigation of *stoichiometry* in the virtual laboratory achieved the same results as the students who conducted research in the laboratory using physical materials and equipment. There were no significant differences between the mean assessment values of the virtual and physical laboratory groups.

In a subsequent study, Olympiou and Zacharia (2012) conducted an experiment with a pre-/post-comparative study methodology in three conditions: 23 students who used PM, 23 students who used VM and 24 students who used a mixed combination of PM and VM. The results indicated that the use of a mixed combination of PM and VM improved the students' conceptual understanding of *light and colour* more than the use of PM or VM in isolation. There could have been a better understanding in terms of student learning if the researchers had used more data sources, particularly sources that focused on the process and not only the final results.

Implicit and Explicit Instruction in Real and Virtual Laboratories

In the study by Jaakkola and Nurmi (2008), we identified limitations that deserve further investigation. For example, we do not know what would have occurred if the students had used real equipment prior to the simulation. To answer this question, Jaakkola et al. (2011) performed a comparison of the learning outcomes of 50 elementary school students who used only simulations (simulation environment, SE) with the results from students who used a simulation in parallel with real circuits (combination environment, CE) in the field of *electricity*; furthermore, they explored how the learning outcomes in these environments are mediated by implicit (procedural guidance only) and explicit (more guidance for the discovery process) teaching. The results demonstrated that the instructional support had the expected effect on their understanding of electrical circuits when they used only the simulation. It is important to note that although explicit teaching was able to considerably improve the students' conceptual understanding of electric circuits in the simulation environment, their understanding did not reach the level of the students in the combined environment. The students had more difficulty when they received more in-depth guidelines in a sequential combination of PM and VM.

Pedagogical Proposals for the Integration of Virtual Laboratories: Case Studies and Constructivist Teaching Strategies

The study of Dori and Sasson (2008) presents a pedagogical proposal that integrates a virtual laboratory with *case studies*. The researchers investigated the chemical understanding and graphing skills of 857 twelfth-grade chemistry students from a number of different secondary schools in a case-based computerised laboratory (CCL). The CCL is a chemistry learning environment that integrates computerised experiments, with an emphasis on scientific research and understanding case studies. The students who used the CCL learning environment significantly improved their graphing skills and retention of chemistry understanding compared with the pre- and post-questionnaires. The CCL's contribution was more noticeable for the experimental students at a low academic level because they benefited more from the combination of visual and textual representations. In light of this study, we realised the benefit of contextualising laboratory activities, which are often neglected by teachers. However, the pedagogical proposal of Russell et al. (2004) suggests that an effective approach to catalyse the construction of the student's understanding may be to connect the power and flexibility of a microcomputer-based laboratory (MBL) with teaching strategies established and based on constructivist learning theories. These strategies primarily depend on the actions of the teacher, who could use guidelines from Barab et al. (2009), Dori and Sasson (2008), Hakkarainen (2003), and other researchers.

Preferences for the Use of Virtual Laboratories

There are studies that investigated student preferences regarding the use of virtual laboratories to study scientific phenomena. For example, Sun et al. (2008) explored the effect of learning about *acids and alcohols* as it related to the different learning styles of 132 primary school students in a Web-based virtual science laboratory. The researchers demonstrated that the students in the experimental group achieved better grades than the control group with the traditional teaching method, the Web-based virtual learning environment is suitable for different learning styles and the simulated experiments promote interest in science learning and enable individualised learning. However, at least 75% of the students surveyed indicated that they preferred to use the Web-based virtual laboratory than to read texts. This preference is consistent with the results of Dori and Belcher (2005), who investigated undergraduate students' preference for active learning. Students who have always had traditional classes find it difficult to accept a more active and less teacher-centred education in their daily lives.

1.4.2.2   Emphasis on Remote Laboratories and Data

One group of studies that has intensified in recent years is related to *remote-controlled experiments*, that is, *real-time computer-based or Internet-based controlled experiments* (Kong et al., 2009; Lowe et al., 2012; Underwood et al., 2008).

Effects of Remotely Controlled Laboratories

Experiments that are remotely controlled via the Internet allow students to manipulate or control real devices to complete experimental activities for scientific research at a distance, using specific hardware and software. The study of Kong et al. (2009) sought to understand student learning outcomes after the use of a remotely controlled system called LabVNC, which is a free software, and the teacher's opinions regarding the use of LabVNC to teach science. The teacher's statement on the pedagogical value of the remotely controlled experiment and the students' enthusiasm for using LabVNC indicate the potential to integrate it in experimental activities. The same results are shown in Lowe et al. (2012), which described trials that used remote laboratories within secondary schools and examined the reactions of 112 students and teachers and their interactions with the laboratories.

The studies demonstrate that when the remote laboratory is used correctly, it can induce a variety of potential benefits, including the ability to share resources and access to equipment across different institutions (which would otherwise be inaccessible because of costs or technical reasons), as well as provide an increase in experimental activity (Kong et al., 2009; Lowe et al., 2012). Although there has been a considerable increase in the use of remote laboratories within higher education, their role in secondary school science teaching remains much more limited.

It is important to note that there are also studies regarding the use of remote research data in collaboration with scientists to be analysed in science teaching. For example, the study of Underwood et al. (2008) presents a framework to identify and describe the resources, tools and services necessary to use *e-Science* in the classroom, to enable local and remote communication and collaboration on scientific topics and with scientific data (Underwood et al., 2008).

1.5 Conclusions

This chapter presented a reflection on the main ICTE (ICT tools and digital resources) and how they have been used to teach science. The main themes are summarised in Table 1.2. Recent applications of the integration of ICT in new pedagogical teaching models should be investigated as curricular reorganisation trends and new forms of science teaching (Dori & Belcher, 2005). It is clear that similar to any other educational tool, the effectiveness of ICT is limited by how they are used (Lee et al., 2011; Ng & Gunstone, 2002; Rutten et al., 2012; Smetana & Bell, 2012; Webb, 2005). Certainly, the participating subjects and the teaching strategies used to support student science learning should be observed in the use of different ICT. In short, the different ICT used to teach science and identified from the literature allowed an understanding of the different ways that ICT-based science content could be developed. Supported by a more innovative science curriculum within a differentiated structure of pedagogical practice, ICT-rich environments provide a range of possibilities that enable science learning. The challenge for teachers is to understand the potential of these resources to support teaching, such as cognitive development, formative evaluation and the development of new science curricula. Using the studies presented here, along with the students' knowledge and a pedagogical knowledge of the content, teachers can negotiate different possibilities for learning science with their students.

From what has been presented previously, we can arrive at the following conclusion:

- The integration of ICT in science education has positive effects when the students' profile, the didactic aspects and the teacher's mediation are taken into consideration.
- The use of ICT is predominately characterised by active processes, which are student-centred rather than teacher-centred.

These initial conclusions are guidelines for future research and will be analysed separately.

1.5.1   The Integration of ICT in Science Teaching Has Positive Effects When the Students' Profile, the Didactic Aspects and the Teacher's Mediation are Taken into Consideration

Many of the studies mentioned presented a wide range of issues and challenges for teachers; however, a key question for us is whether the use of ICT requires significant changes to the teacher's role or a "new pedagogy" compared with other teaching strategies (see, e.g. Athanassios, 2010; Barton, 2005; De Winter et al., 2010). ICT clearly have the potential to support science teaching; however, these considerations raise new challenges for us as educators: does an ICT pedagogy exist, or is an ICT pedagogy necessary to teach science? First, we determined that in most of the studies, the results did not take into account the potential impact of teacher support, the classroom setting or the place of ICT within the curriculum (Rutten et al., 2012; Scalise et al., 2011; Smetana & Bell, 2012). The learning outcomes, evolution and conceptual understanding promoted by ICT in the classroom differ in many ways, including the target audience, their objective, the use of technological tools, the teacher's autonomy and the flexibility of the methodology for data collection. The effective or ineffective use of these resources for teaching science will also often be related to the teacher's actions; they are not always prepared or have had proper training to use ICT in the classroom (Barak & Dori, 2011; Kumar et al., 2011; Yarden & Yarden, 2011). Second, we determined that the positive results of different studies only occurred because the teacher had a role in guiding many actions (Hakkarainen, 2003; Ng & Gunstone, 2002).

1.5.2   The Use of ICT Is Predominately Characterised by Active Processes, Which Are Student-Centred Rather than Teacher-Centred

We can summarise the main reflections of this chapter in the four main effects identified in the work of Webb (2005), which support ICT-based science learning: (1) promote cognitive acceleration; (2) enable a wide range of experiences so that students can relate science with their experiences and other experiences of the real world; (3) increase student self-management; and (4) facilitate data collection and presentation. These effects only make sense if we consider that ICT empower active, student-centred processes, even though there is a need for teacher mediation in many processes, as well as in the use of many ICT resources.

We determined that it is not always easy to create student-centred conditions in which the teacher ceases to be an active agent and becomes the manager of the learning process. The work of Dori and Belcher (2005) is an example of this situation; the study analysed the effects of a media-rich differentiated learning space, the TEAL project, with students at MIT. The project's objective was to break with traditional teaching, thereby making the student more active in their learning process. The results indicated that the students on the small-scale, who studied in the TEAL format, significantly improved their conceptual understanding of the phenomena studied compared with the students on the large-scale. The positive attitudes were related to knowledge gains, and the negative attitudes were related to the acceptance of the project; in these types of basic physics courses, students have traditionally been used to lecture-based classes, having passive attitudes, closely following a textbook and studying for exams. Thus, making the student become active in their learning process should begin during their early years and continue through the entire educational process, which remains distant for many schools and teacher training centres.

References

Anastopoulou, S., Sharples, M., & Baber, C. (2011). An evaluation of multimodal interactions with technology while learning science concepts. *British Journal of Educational Technology, 42*(2), 266–290. https://doi.org/10.1111/j.1467-8535.2009.01017.x

Ardac, D., & Akaygun, S. (2004). Effectiveness of multimedia-based instruction that emphasizes molecular representations on students' understanding of chemical change. *Journal of Research in Science Teaching, 41*, 317–337. https://doi.org/10.1002/tea.20005

Athanassios, J. (2010). Designing and implementing an integrated technological pedagogical science knowledge framework for science teachers professional development. *Computers & Education, 55*, 1259–1269. https://doi.org/10.1016/j.compedu.2010.05.022

Barab, S., & Dede, C. (2007). Games and immersive participatory simulations for science education: An emerging type of curricula. *Journal of Science Education and Technology, 16*, 1–3. https://doi.org/10.1007/s10956-007-9043-9

Barab, S., Sadler, T., Heiselt, C., Hickey, D., & Zuiker, S. (2007). Relating narrative, inquiry, and inscriptions: Supporting consequential play. *Journal of Science Education and Technology, 16*, 59–82. https://doi.org/10.1007/s10956-006-9033-3

Barab, S., Scott, B., Siyahhan, S., Goldstone, R., Ingram-Goble, A., Zuiker, S., & Warren, S. (2009). Transformational play as a curricular scaffold: Using videogames to support science education. *Journal of Science Education and Technology, 18*, 305–320. https://doi.org/10.1007/s10956-009-9171-5

Barak, M., Ashkar, T., & Dori, Y. J. (2011). Learning science via animated movies: Its effect on students' thinking and motivation. *Computers & Education, 56*, 839–846. https://doi.org/10.1016/j.compedu.2010.10.025

Barak, M., & Dori, Y. (2011). Science education in primary schools: Is an animation worth a thousand pictures? *Journal of Science Education and Technology, 20*, 608–620. https://doi.org/10.1007/s10956-011-9315-2

Barak, M., & Dori, Y. J. (2005). Enhancing undergraduate students' chemistry understanding through project-based learning in an IT environment. *Science Education, 89*, 117–139. https://doi.org/10.1002/sce.20027

Barton, R. (2005). Supporting teachers in making innovative changes in the use of computer-aided practical work to support concept development in physics education. *International Journal of Science Education, 27*, 345–365. https://doi.org/10.1080/0950069042000230794

Bell, R., & Bell, L. (2003). A bibliography of articles on instructional technology in science education. *Contemporary Issues in Technology and Teacher Education, 2*(4). Retrieved from http://www.citejournal.org/vol2/iss4/science/article2.cfm

Byrne, J., & Grace, M. (2010). Using a concept mapping tool with a photograph association technique (compat) to elicit children's ideas about microbial activity. *International Journal of Science Education, 32*, 479–500. https://doi.org/10.1080/09500690802688071

Chang, C., & Tsai, C. (2005). The interplay between different forms of CAI and students' preferences of learning environment in the secondary science class. *Science Education, 89*, 707–724. https://doi.org/10.1002/sce.20072

Chang, C., Yeh, T., & Barufaldi, J. P. (2010). The positive and negative effects of science concept tests on student conceptual understanding. *International Journal of Science Education, 32*, 265–282. https://doi.org/10.1080/09500690802650055

Chang, C.-Y., Hsiao, C.-H., & Chang, Y.-H. (2011). Science learning outcomes in alignment with learning environment preferences. *Journal of Science Education and Technology, 20*, 136–145. https://doi.org/10.1007/s10956-010-9240-9

Chang, H.-Y., Quintana, C., & Krajcik, J. S. (2010). The impact of designing and evaluating molecular animations on how well middle school students understand the particulate nature of matter. *Science Education, 94*, 73–94. https://doi.org/10.1002/sce.20352

Charlier, B., Peraya, D., & Collectif. (2007). *Transformation des regards sur la recherche en technologie de l'éducation*. Bruxelles, Belgium: De Boeck.

Chen, W., & Looi, C.-K. (2011). Active classroom participation in a group scribbles primary science classroom. *British Journal of Educational Technology, 42*, 676–686. https://doi.org/10.1111/j.1467-8535.2010.01082.x

Chen, Y.-H., Jang, S.-J., & Chen, P.-J. (2015). Using wikis and collaborative learning for science teachers' professional development. *Journal of Computer Assisted Learning, 31*(4), 330–344. https://doi.org/10.1111/jcal.12095

Cher Ping, L. (2008). Global citizenship education, school curriculum and games: Learning mathematics, English and science as a global citizen. *Computers & Education, 51*, 1073–1093. https://doi.org/10.1016/j.compedu.2007.10.005

Chin-Chung, T. (2009). Conceptions of learning versus conceptions of web-based learning: The differences revealed by college students. *Computers & Education, 53*(4), 1092–1103. https://doi.org/10.1016/j.compedu.2009.05.019

Clark, D., & Jorde, D. (2004). Helping students revise disruptive experientially supported ideas about thermodynamics: Computer visualizations and tactile models. *Journal of Research in Science Teaching, 41*, 1–23. https://doi.org/10.1002/tea.10097

Clark, D., & Sampson, V. D. (2007). Personally-seeded discussions to scaffold online argumentation. *International Journal of Science Education, 29*, 253–277. https://doi.org/10.1080/09500690600560944

Connolly, T. M., Boyle, E. A., MacArthur, E., Hainey, T., & Boyle, J. M. (2012). A systematic literature review of empirical evidence on computer games and serious games. *Computers & Education, 59*, 661–686. https://doi.org/10.1016/j.compedu.2012.03.004

Dalacosta, K., Kamariotaki-Paparrigopoulou, M., Palyvos, J. A., & Spyrellis, N. (2009). Multimedia application with animated cartoons for teaching science in elementary education. *Computers & Education, 52*, 741–748. https://doi.org/10.1016/j.compedu.2008.11.018

De Winter, J., Winterbottom, M., & Wilson, E. (2010). Developing a user guide to integrating new technologies in science teaching and learning: Teachers' and pupils' perceptions of their affordances. *Technology, Pedagogy and Education, 19*, 261–267. https://doi.org/10.1080/1475939X.2010.491237

Dede, C., & Barab, S. (2009). Emerging technologies for learning science: A time of rapid advances. *Journal of Science Education and Technology, 18*, 301–304. https://doi.org/10.1007/s10956-009-9172-4

Devolder, A., van Braak, J., & Tondeur, J. (2012). Supporting self-regulated learning in computer-based learning environments: Systematic review of effects of scaffolding in the domain of science education. *Journal of Computer Assisted Learning, 28*, 557–573. https://doi.org/10.1111/j.1365-2729.2011.00476.x

Dimopoulos, K., & Asimakopoulos, A. (2009). Science on the web: Secondary school students' navigation patterns and preferred pages' characteristics. *Journal of Science Education and Technology, 19*, 246–265. https://doi.org/10.1007/s10956-009-9197-8

Dimopoulos, K., & Asimakopoulos, A. (2010). Science on the web: Secondary school students' navigation patterns and preferred pages' characteristics. *Journal of Science Education and Technology, 19,* 246–265. https://doi.org/10.1007/s10956-009-9197-8

Donnelly, D. F., & Boniface, S. (2013). Consuming and creating: Early-adopting science teachers' perceptions and use of a wiki to support professional development. *Computers & Education, 68,* 9–20. https://doi.org/10.1016/j.compedu.2013.04.023

Dori, Y. J., & Belcher, J. (2005). How does technology-enabled active learning affect undergraduate students' understanding of electromagnetism concepts? *Journal of the Learning Sciences, 14,* 243–279. https://doi.org/10.1207/s15327809jls1402_3

Dori, Y. J., & Sasson, I. (2008). Chemical understanding and graphing skills in an honors case-based computerized chemistry laboratory environment: The value of bidirectional visual and textual representations. *Journal of Research in Science Teaching, 45,* 219–250. https://doi.org/10.1002/tea.20197

Ebenezer, J., Kaya, O. N., & Ebenezer, D. L. (2011). Engaging students in environmental research projects: Perceptions of fluency with innovative technologies and levels of scientific inquiry abilities. *Journal of Research in Science Teaching, 48,* 94–116. https://doi.org/10.1002/tea.20387

Ekanayake, S. Y., & Wishart, J. (2015). Integrating mobile phones into teaching and learning: A case study of teacher training through professional development workshops. *British Journal of Educational Technology, 46*(1), 173–189. https://doi.org/10.1111/bjet.12131

Enyedy, N., Danish, J., Delacruz, G., & Kumar, M. (2012). Learning physics through play in an augmented reality environment. *International Journal of Computer-Supported Collaborative Learning, 7,* 347–378. https://doi.org/10.1007/s11412-012-9150-3

Ergazaki, M., Zogza, V., & Komis, V. (2007). Analysing students' shared activity while modeling a biological process in a computer-supported educational environment. *Journal of Computer Assisted Learning, 23,* 158–168. https://doi.org/10.1111/j.1365-2729.2006.00214.x

Evagorou, M., Korfiatis, K., Nicolaou, C., & Constantinou, C. (2009). An investigation of the potential of interactive simulations for developing system thinking skills in elementary school: A case study with fifth-graders and sixth-graders. *International Journal of Science Education, 31,* 655–674. https://doi.org/10.1080/09500690701749313

Furberg, A., & Ludvigsen, S. (2008). Students' meaning-making of socio-scientific issues in computer mediated settings: Exploring learning through interaction trajectories. *International Journal of Science Education, 30,* 1775–1799. https://doi.org/10.1080/09500690701543617

Furman, M., & Barton, A. C. (2006). Capturing urban student voices in the creation of a science mini-documentary. *Journal of Research in Science Teaching, 43,* 667–694. https://doi.org/10.1002/tea.20164

Gelbart, H., Brill, G., & Yarden, A. (2009). The impact of a web-based research simulation in bioinformatics on students' understanding of genetics. *Research in Science Education, 39,* 725–751. https://doi.org/10.1007/s11165-008-9101-1

Hakkarainen, K. (2003). Progressive inquiry in a computer-supported biology class. *Journal of Research in Science Teaching, 40,* 1072–1088. https://doi.org/10.1002/tea.10121

Hansen, J. A., Barnett, M., MaKinster, J. G., & Keating, T. (2004). The impact of three-dimensional computational modeling on student understanding of astronomical concepts: A quantitative analysis. *International Journal of Science Education, 26,* 1365–1378. https://doi.org/10.1080/09500690420001673757

Hansson, L., Redfors, A., & Rosberg, M. (2011). Students' socio-scientific reasoning in an astrobiological context during work with a digital learning environment. *Journal of Science Education and Technology, 20,* 388–402. https://doi.org/10.1007/s10956-010-9260-5

Höffler, T. N., & Leutner, D. (2007). Instructional animation versus static pictures: A meta-analysis. *Learning and Instruction, 17,* 722–738. https://doi.org/10.1016/j.learninstruc.2007.09.013

Hoffman, J. L., Wu, H.-K., Krajcik, J. S., & Soloway, E. (2003). The nature of middle school learners' science content understandings with the use of on-line resources. *Journal of Research in Science Teaching, 40,* 323–346. https://doi.org/10.1002/tea.10079

Holbrook, J., & Dupont, C. (2011). Making the decision to provide enhanced podcasts to post-secondary science students. *Journal of Science Education and Technology, 20,* 233–245. https://doi.org/10.1007/s10956-010-9248-1

Howard, S. K., Chan, A., & Caputi, P. (2015). More than beliefs: Subject areas and teachers' integration of laptops in secondary teaching. *British Journal of Educational Technology, 46*(2), 360–369. https://doi.org/10.1111/bjet.12139

Hsu, C.-Y., Tsai, C.-C., & Liang, J.-C. (2011). Facilitating preschoolers' scientific knowledge construction via computer games regarding light and shadow: The effect of the prediction-observation-explanation (POE) strategy. *Journal of Science Education and Technology, 20*, 482–493. https://doi.org/10.1007/s10956-011-9298-z

Hsu, Y.-S. (2006). Lesson rainbow: The use of multiple representations in an Internet-based, discipline-integrated science lesson. *British Journal of Educational Technology, 37*, 539–557. https://doi.org/10.1111/j.1467-8535.2006.00551.x

Hsu, Y.-S., Wu, H.-K., & Hwang, F.-K. (2008). Fostering high school students' conceptual understandings about seasons: The design of a technology-enhanced learning environment. *Research in Science Education, 38*, 127–147. https://doi.org/10.1007/s11165-007-9041-1

Jaakkola, T., & Nurmi, S. (2008). Fostering elementary school students' understanding of simple electricity by combining simulation and laboratory activities. *Journal of Computer Assisted Learning, 24*, 271–283. https://doi.org/10.1111/j.1365-2729.2007.00259.x

Jaakkola, T., Nurmi, S., & Veermans, K. (2011). A comparison of students' conceptual understanding of electric circuits in simulation only and simulation-laboratory contexts. *Journal of Research in Science Teaching, 48*, 71–93. https://doi.org/10.1002/tea.20386

Jacobson, M. J., & Archodidou, A. (2000). The design of hypermedia tools for learning: Fostering conceptual change and transfer of complex scientific knowledge. *Journal of the Learning Sciences, 9*, 145–199. https://doi.org/10.1207/s15327809jls0902_2

Jang, S. (2006). The effects of incorporating web-assisted learning with team teaching in seventh-grade science classes. *International Journal of Science Education, 28*, 615–632. https://doi.org/10.1080/09500690500339753

Katz, P. (2011). A case study of the use of internet photobook technology to enhance early childhood "scientist" identity. *Journal of Science Education and Technology, 20*, 525–536. https://doi.org/10.1007/s10956-011-9301-8

Ketelhut, D. (2007). The impact of student self-efficacy on scientific inquiry skills: An exploratory investigation in river city, a multi-user virtual environment. *Journal of Science Education and Technology, 16*, 99–111. https://doi.org/10.1007/s10956-006-9038-y

Khan, S. (2010). New pedagogies on teaching science with computer simulations. *Journal of Science Education and Technology, 20*(3), 215–232. https://doi.org/10.1007/s10956-010-9247-2

Khan, S. (2011). New pedagogies on teaching science with computer simulations. *Journal of Science Education and Technology, 20*, 215–232. https://doi.org/10.1007/s10956-010-9247-2

Kim, H., & Herbert, B. (2012). Inquiry resources collection as a boundary object supporting meaningful collaboration in a wiki-based scientist-teacher community. *Journal of Science Education and Technology, 21*, 504–512. https://doi.org/10.1007/s10956-011-9342-z

Kim, H., Miller, H., Herbert, B., Pedersen, S., & Loving, C. (2012). Using a wiki in a scientist-teacher professional learning community: Impact on teacher perception changes. *Journal of Science Education and Technology, 21*(4), 440–452. https://doi.org/10.1007/s10956-011-9336-x

Klisch, Y., Miller, L., Wang, S., & Epstein, J. (2012). The impact of a science education game on students' learning and perception of inhalants as body pollutants. *Journal of Science Education and Technology, 21*, 295–303. https://doi.org/10.1007/s10956-011-9319-y

Kong, S. C., Yeung, Y. Y., & Wu, X. Q. (2009). An experience of teaching for learning by observation: Remote-controlled experiments on electrical circuits. *Computers & Education, 52*, 702–717. https://doi.org/10.1016/j.compedu.2008.11.011

Kubasko, D., Jones, M. G., Tretter, T., & Andre, T. (2008). Is it live or is it memorex? Students' synchronous and asynchronous communication with scientists. *International Journal of Science Education, 30*, 495–514. https://doi.org/10.1080/09500690701217220

Kumar, D., Thomas, P., Morris, J., Tobias, K., Baker, M., & Jermanovich, T. (2011). Effect of current electricity simulation supported learning on the conceptual understanding of elementary and secondary teachers. *Journal of Science Education and Technology, 20,* 111–115. https://doi.org/10.1007/s10956-010-9229-4

Lavonen, J., Juuti, K., & Meisalo, V. (2003). Designing a user-friendly microcomputer-based laboratory package through the factor analysis of teacher evaluations. *International Journal of Science Education, 25*(12), 1471–1487. https://doi.org/10.1080/0950069032000072755

Lee, S. W., Tsai, C., Wu, Y., Tsai, M., Liu, T., Hwang, F., et al. (2011). Internet-based science learning: A review of journal publications. *International Journal of Science Education, 33,* 1893–1925. https://doi.org/10.1080/09500693.2010.536998

Li, S. C., Law, N., & Lui, K. F. A. (2006). Cognitive perturbation through dynamic modelling: A pedagogical approach to conceptual change in science. *Journal of Computer Assisted Learning, 22,* 405–422. https://doi.org/10.1111/j.1365-2729.2006.00187.x

Lim, C. P., Nonis, D., & Hedberg, J. (2006). Gaming in a 3D multiuser virtual environment: Engaging students in science lessons. *British Journal of Educational Technology, 37,* 211–231. https://doi.org/10.1111/j.1467-8535.2006.00531.x

Lin, L.-F., Hsu, Y.-S., & Yeh, Y.-F. (2012). The role of computer simulation in an inquiry-based learning environment: Reconstructing geological events as geologists. *Journal of Science Education and Technology, 21,* 370–383. https://doi.org/10.1007/s10956-011-9330-3

Ling Wong, S., Wai Yung, B. H., Cheng, M. W., Lam, K. L., & Hodson, D. (2006). Setting the Stage for Developing Pre-service Teachers' Conceptions of Good Science Teaching: The role of classroom videos. *International Journal of Science Education, 28*(1), 1–24. https://doi.org/10.1080/09500690500239805

Lindgren, R., & Schwartz, D. L. (2009). Spatial learning and computer simulations in science. *International Journal of Science Education, 31,* 419–438. https://doi.org/10.1080/09500690802595813

Liu, L., & Hmelo-Silver, C. E. (2009). Promoting complex systems learning through the use of conceptual representations in hypermedia. *Journal of Research in Science Teaching, 46,* 1023–1040. https://doi.org/10.1002/tea.20297

Looi, C.-K., Zhang, B., Chen, W., Seow, P., Chia, G., Norris, C., & Soloway, E. (2011). 1:1 mobile inquiry learning experience for primary science students: A study of learning effectiveness. *Journal of Computer Assisted Learning, 27,* 269–287. https://doi.org/10.1111/j.1365-2729.2010.00390.x

Lowe, D., Newcombe, P., & Stumpers, B. (2012). Evaluation of the use of remote laboratories for secondary school science education. *Research in Science Education, 43,* 1–23. https://doi.org/10.1007/s11165-012-9304-3

Marbach-Ad, G., Rotbain, Y., & Stavy, R. (2008). Using computer animation and illustration activities to improve high school students' achievement in molecular genetics. *Journal of Research in Science Teaching, 45,* 273–292. https://doi.org/10.1002/tea.20222

Mayer-Smith, J., Pedretti, E., & Woodrow, J. (2000). Closing of the gender gap in technology enriched science education: A case study. *Computers & Education, 35,* 51–63. https://doi.org/10.1016/S0360-1315(00)00018-X

Mayo, A., Sharma, M., & Muller, D. (2009). Qualitative differences between learning environments using videos in small groups and whole class discussions: A preliminary study in physics. *Research in Science Education, 39,* 477–493. https://doi.org/10.1007/s11165-008-9090-0

McConnell, T. J., Parker, J. M., Eberhardt, J., Koehler, M. J., & Lundeberg, M. A. (2012). Virtual Professional Learning Communities: Teachers' Perceptions of Virtual Versus Face-to-Face Professional Development. *Journal of Science Education and Technology, 22*(3), 267–277. https://doi.org/10.1007/s10956-012-9391-y

Mistler-Jackson, M., & Butler Songer, N. (2000). Student motivation and internet technology: Are students empowered to learn science? *Journal of Research in Science Teaching, 37,* 459–479. https://doi.org/10.1002/(SICI)1098-2736(200005)37:5<459::AID-TEA5>3.0.CO;2-C

Moss, K., & Crowley, M. (2011). Effective learning in science: The use of personal response systems with a wide range of audiences. *Computers & Education, 56,* 36–43. https://doi.org/10.1016/j.compedu.2010.03.021

Nelson, B. (2007). Exploring the use of individualized, reflective guidance in an educational multi-user virtual environment. *Journal of Science Education and Technology, 16*, 83–97. https://doi.org/10.1007/s10956-006-9039-x

Neulight, N., Kafai, Y., Kao, L., Foley, B., & Galas, C. (2007). Children's participation in a virtual epidemic in the science classroom: Making connections to natural infectious diseases. *Journal of Science Education and Technology, 16*, 47–58. https://doi.org/10.1007/s10956-006-9029-z

Ng, W., & Gunstone, R. (2002). Students' perceptions of the effectiveness of the world wide web as a research and teaching tool in science learning. *Research in Science Education, 32*, 489–510. https://doi.org/10.1023/A:1022429900836

Nielsen, W., Miller, K. A., & Hoban, G. (2014). Science Teachers' Response to the Digital Education Revolution. *Journal of Science Education and Technology, 24*(4), 417–431. https://doi.org/10.1007/s10956-014-9527-3

Olde, C. V., & de Jong, T. (2004). Student-generated assignments about electrical circuits in a computer simulation. *International Journal of Science Education, 26*, 859–873. https://doi.org/10.1080/0950069032000138815

Olympiou, G., & Zacharia, Z. C. (2012). Blending physical and virtual manipulatives: An effort to improve students' conceptual understanding through science laboratory experimentation. *Science Education, 96*, 21–47. https://doi.org/10.1002/sce.20463

Orion, N., Dubowski, Y., & Dodick, J. (2000). The educational potential of multimedia authoring as a part of the earth science curriculum—A case study. *Journal of Research in Science Teaching, 37*, 1121–1153. https://doi.org/10.1002/1098-2736(200012)37:10<1121::AID-TEA6>3.0.CO;2-L

Oshima, J., Oshima, R., Murayama, I., Inagaki, S., Takenaka, M., Nakayama, H., & Yamaguchi, E. (2004). Design experiments in Japanese elementary science education with computer support for collaborative learning: Hypothesis testing and collaborative construction. *International Journal of Science Education, 26*, 1199–1221. https://doi.org/10.1080/0950069032000138824

Park, H., Khan, S., & Petrina, S. (2009). ICT in science education: A quasi-experimental study of achievement, attitudes toward science, and career aspirations of Korean middle school students. *International Journal of Science Education, 31*, 993–1012. https://doi.org/10.1080/09500690701787891

Pata, K., & Sarapuu, T. (2006). A comparison of reasoning processes in a collaborative modelling environment: Learning about genetics problems using virtual chat. *International Journal of Science Education, 28*, 1347–1368. https://doi.org/10.1080/09500690500438670

Pedretti, E., Mayer-Smith, J., & Woodrow, J. (1998). Technology, text, and talk: Students' perspectives on teaching and learning in a technology-enhanced secondary science classroom. *Science Education, 82*, 569–589. https://doi.org/10.1002/(SICI)1098-237X(199809)82:5<569::AID-SCE3>3.0.CO;2-7

Piburn, M. D., Reynolds, S. J., McAuliffe, C., Leedy, D. E., Birk, J. P., & Johnson, J. K. (2005). The role of visualization in learning from computer-based images. *International Journal of Science Education, 27*, 513–527. https://doi.org/10.1080/09500690412331314478

Plass, J. L., Milne, C., Homer, B. D., Schwartz, R. N., Hayward, E. O., Jordan, T., et al. (2012). Investigating the effectiveness of computer simulations for chemistry learning. *Journal of Research in Science Teaching, 49*, 394–419. https://doi.org/10.1002/tea.21008

Pol, H., Harskamp, E., & Suhre, C. (2005). Solving physics problems with the help of computer-assisted instruction. *International Journal of Science Education, 27*, 451–469. https://doi.org/10.1080/0950069042000266164

Pombo, L., Smith, M., Abelha, M., Caixinha, H., & Costa, N. (2012). Evaluating an online e-module for Portuguese primary teachers: trainees' perceptions. *Technology, Pedagogy and Education, 21*(1), 21–36. https://doi.org/10.1080/1475939X.2011.589156

Price, S., Davies, P., Farr, W., Jewitt, C., Roussos, G., & Sin, G. (2013). Fostering geospatial thinking in science education through a customisable smartphone application. *British Journal of Educational Technology, 45*(1), 160–170. https://doi.org/10.1111/bjet.12000

Pyatt, K., & Sims, R. (2012). Virtual and physical experimentation in inquiry-based science labs: Attitudes, performance and access. *Journal of Science Education and Technology, 21*, 133–147. https://doi.org/10.1007/s10956-011-9291-6

Quellmalz, E. S., Timms, M. J., Silberglitt, M. D., & Buckley, B. C. (2012). Science assessments for all: Integrating science simulations into balanced state science assessment systems. *Journal of Research in Science Teaching, 49*, 363–393. https://doi.org/10.1002/tea.21005

Reid, D. J., Zhang, J., & Chen, Q. (2003). Supporting scientific discovery learning in a simulation environment. *Journal of Computer Assisted Learning, 19*, 9–20. https://doi.org/10.1046/j.0266-4909.2003.00002.x

Ronen, M., & Eliahu, M. (1999). Simulation as a home learning environment — students' views. *Journal of Computer Assisted Learning, 15*, 258–268. https://doi.org/10.1046/j.1365-2729.1999.00101.x

Ronen, M., & Eliahu, M. (2000). Simulation — a bridge between theory and reality: The case of electric circuits. *Journal of Computer Assisted Learning, 16*, 14–26. https://doi.org/10.1046/j.1365-2729.2000.00112.x

Rosenbaum, E., Klopfer, E., & Perry, J. (2007). On location learning: Authentic applied science with networked augmented realities. *Journal of Science Education and Technology, 16*, 31–45. https://doi.org/10.1007/s10956-006-9036-0

Roth, K. J., Garnier, H. E., Chen, C., Lemmens, M., Schwille, K., & Wickler, N. I. Z. (2011). Videobased lesson analysis: Effective science PD for teacher and student learning. *Journal of Research in Science Teaching, 48*(2), 117–148. https://doi.org/10.1002/tea.20408

Russell, D. W., Lucas, K. B., & McRobbie, C. J. (2004). Role of the microcomputer-based laboratory display in supporting the construction of new understandings in thermal physics. *Journal of Research in Science Teaching, 41*, 165–185. https://doi.org/10.1002/tea.10129

Rutten, N., van Joolingen, W. R., & van der Veen, J. T. (2012). The learning effects of computer simulations in science education. *Computers & Education, 58*, 136–153. https://doi.org/10.1016/j.compedu.2011.07.017

Şad, S. N., & Göktaş, Ö. (2014). Preservice teachers' perceptions about using mobile phones and laptops in education as mobile learning tools. *British Journal of Educational Technology, 45*(4), 606–618. https://doi.org/10.1111/bjet.12064

Scalise, K. (2012). Science learning and instruction: Taking advantage of technology to promote knowledge integration. *Science Education, 96*, 1136–1138. https://doi.org/10.1002/sce.21025

Scalise, K., Timms, M., Moorjani, A., Clark, L., Holtermann, K., & Irvin, P. S. (2011). Student learning in science simulations: Design features that promote learning gains. *Journal of Research in Science Teaching, 48*, 1050–1078. https://doi.org/10.1002/tea.20437

Schaal, S., Bogner, F., & Girwidz, R. (2010). Concept mapping assessment of media assisted learning in interdisciplinary science education. *Research in Science Education, 40*, 339–352. https://doi.org/10.1007/s11165-009-9123-3

Shapiro, A. M. (1999). The relevance of hierarchies to learning biology from hypertext. *Journal of the Learning Sciences, 8*, 215–243. https://doi.org/10.1207/s15327809jls0802_2

She, H.-C., & Chen, Y.-Z. (2009). The impact of multimedia effect on science learning: Evidence from eye movements. *Computers & Education, 53*, 1297–1307. https://doi.org/10.1016/j.compedu.2009.06.012

She, H.-C., Cheng, M.-T., Li, T.-W., Wang, C.-Y., Chiu, H.-T., Lee, P.-Z., et al. (2012). Web-based undergraduate chemistry problem-solving: The interplay of task performance, domain knowledge and web-searching strategies. *Computers & Education, 59*, 750–761. https://doi.org/10.1016/j.compedu.2012.02.005

She, H.-C., & Lee, C.-Q. (2008). SCCR digital learning system for scientific conceptual change and scientific reasoning. *Computers & Education, 51*, 724–742. https://doi.org/10.1016/j.compedu.2007.07.009

She, H.-C., & Liao, Y.-W. (2010). Bridging scientific reasoning and conceptual change through adaptive web-based learning. *Journal of Research in Science Teaching, 47*, 91–119. https://doi.org/10.1002/tea.20309

Shin, N., Jonassen, D. H., & McGee, S. (2003). Predictors of well-structured and ill-structured problem solving in an astronomy simulation. *Journal of Research in Science Teaching, 40*, 6–33. https://doi.org/10.1002/tea.10058

Smetana, L. K., & Bell, R. L. (2012). Computer simulations to support science instruction and learning: A critical review of the literature. *International Journal of Science Education, 34*, 1337–1370. https://doi.org/10.1080/09500693.2011.605182

Soong, B., & Mercer, N. (2011). Improving students' revision of physics concepts through ICT-based co-construction and prescriptive tutoring. *International Journal of Science Education, 33*(8), 1055–1078. https://doi.org/10.1080/09500693.2010.489586

Squire, K., & Jan, M. (2007). Mad city mystery: Developing scientific argumentation skills with a place-based augmented reality game on handheld computers. *Journal of Science Education and Technology, 16*, 5–29. https://doi.org/10.1007/s10956-006-9037-z

Starbek, P., Starčič Erjavec, M., & Peklaj, C. (2010). Teaching genetics with multimedia results in better acquisition of knowledge and improvement in comprehension. *Journal of Computer Assisted Learning, 26*, 214–224. https://doi.org/10.1111/j.1365-2729.2009.00344.x

Stieff, M. (2011). Improving representational competence using molecular simulations embedded in inquiry activities. *Journal of Research in Science Teaching, 48*, 1137–1158. https://doi.org/10.1002/tea.20438

Strømme, T. A., & Furberg, A. (2015). Exploring Teacher Intervention in the Intersection of Digital Resources, Peer Collaboration, and Instructional Design. *Science Education, 99*(5), 837–862. https://doi.org/10.1002/sce.21181

Sun, K., Lin, Y., & Yu, C. (2008). A study on learning effect among different learning styles in a web-based lab of science for elementary school students. *Computers & Education, 50*, 1411–1422. https://doi.org/10.1016/j.compedu.2007.01.003

Tekos, G., & Solomonidou, C. (2009). Constructivist learning and teaching of optics concepts using ICT tools in Greek primary school: A pilot study. *Journal of Science Education and Technology, 18*, 415–428. https://doi.org/10.1007/s10956-009-9158-2

Tolentino, L., Birchfield, D., Megowan-Romanowicz, C., Johnson-Glenberg, M. C., Kelliher, A., & Martinez, C. (2009). Teaching and learning in the mixed-reality science classroom. *Journal of Science Education and Technology, 18*, 501–517. https://doi.org/10.1007/s10956-009-9166-2

Tseng, C., Tuan, H., & Chin, C. (2010). Investigating the influence of motivational factors on conceptual change in a digital learning context using the dual-situated learning model. *International Journal of Science Education, 32*, 1853–1875. https://doi.org/10.1080/09500690903219156

Underwood, J., Smith, H., Luckin, R., & Fitzpatrick, G. (2008). E-Science in the classroom – towards viability. *Computers & Education, 50*, 535–546. https://doi.org/10.1016/j.compedu.2007.07.003

Valtonen, T., Hacklin, S., Kontkanen, S., Hartikainen-Ahia, A., Kärkkäinen, S., & Kukkonen, J. (2013). Pre-service teachers' experiences of using social software applications for collaborative inquiry. *Computers & Education, 69*, 85–95. https://doi.org/10.1016/j.compedu.2013.07.001

Veermans, K., van Joolingen, W., & de Jong, T. (2006). Use of heuristics to facilitate scientific discovery learning in a simulation learning environment in a physics domain. *International Journal of Science Education, 28*, 341–361. https://doi.org/10.1080/09500690500277615

Waight, N., & Abd-El-Khalick, F. (2007). The impact of technology on the enactment of "inquiry" in a technology enthusiast's sixth grade science classroom. *Journal of Research in Science Teaching, 44*, 154–182. https://doi.org/10.1002/tea.20158

Wang, C., Ke, Y.-T., Wu, J.-T., & Hsu, W.-H. (2012). Collaborative action research on technology integration for science learning. *Journal of Science Education and Technology, 21*, 125–132. https://doi.org/10.1007/s10956-011-9289-0

Warwick, P., Mercer, N., Kershner, R., & Staarman, J. K. (2010). In the mind and in the technology: The vicarious presence of the teacher in pupil's learning of science in collaborative group activity at the interactive whiteboard. *Computers & Education, 55*, 350–362. https://doi.org/10.1016/j.compedu.2010.02.001

Webb, M. E. (2005). Affordances of ICT in science learning: Implications for an integrated pedagogy. *International Journal of Science Education, 27*, 705–735. https://doi.org/10.1080/09500690500038520

Yarden, H., & Yarden, A. (2011). Studying biotechnological methods using animations: The teacher's role. *Journal of Science Education and Technology, 20*, 689–702. https://doi.org/10.1007/s10956-010-9262-3

Zacharia, Z. C. (2007). Comparing and combining real and virtual experimentation: An effort to enhance students' conceptual understanding of electric circuits. *Journal of Computer Assisted Learning, 23*, 120–132. https://doi.org/10.1111/j.1365-2729.2006.00215.x

Zacharia, Z. C., Olympiou, G., & Papaevripidou, M. (2008). Effects of experimenting with physical and virtual manipulatives on students' conceptual understanding in heat and temperature. *Journal of Research in Science Teaching, 45*, 1021–1035. https://doi.org/10.1002/tea.20260

Zhang, B., Looi, C.-K., Seow, P., Chia, G., Wong, L.-H., Chen, W., et al. (2010). Deconstructing and reconstructing: Transforming primary science learning via a mobilized curriculum. *Computers & Education, 55*, 1504–1523. https://doi.org/10.1016/j.compedu.2010.06.016

Zhang, J., Chen, Q., Sun, Y., & Reid, D. J. (2004). Triple scheme of learning support design for scientific discovery learning based on computer simulation: Experimental research. *Journal of Computer Assisted Learning, 20*, 269–282. https://doi.org/10.1111/j.1365-2729.2004.00062.x

Zhang, M. (2012). Supporting middle school students' online reading of scientific resources: Moving beyond cursory, fragmented, and opportunistic reading. *Journal of Computer Assisted Learning, 29*, 138–152. https://doi.org/10.1111/j.1365-2729.2012.00478.x

Zheng, R. Z., Yang, W., Garcia, D., & McCadden, E. P. (2008). Effects of multimedia and schema induced analogical reasoning on science learning. *Journal of Computer Assisted Learning, 24*, 474–482. https://doi.org/10.1111/j.1365-2729.2008.00282.x

Zydney, J., & Grincewicz, A. (2011). The use of video cases in a multimedia learning environment for facilitating high school students' inquiry into a problem from varying perspectives. *Journal of Science Education and Technology, 20*, 715–728. https://doi.org/10.1007/s10956-010-9264-1

Chapter 2
Different Theoretical Approaches to the Use of ICT in Science Education

2.1 Introduction

The introduction of various digital technologies into the science curriculum has been gaining attention in recent years (Devolder, van Braak, & Tondeur, 2012; Lee et al., 2011; Smetana & Bell, 2012); however, the effective use of these technologies requires well-designed resources, in addition to changes in the educational approaches of teachers (Hsu, Wu, & Hwang, 2008). The number of studies on the effectiveness of technologies and how to introduce them into the science curriculum has been increasing annually; however, little is known about the theoretical approaches that support the production of such resources and their use in the classroom.

Due to the importance of integrating Information and Communication Technologies (ICT) in an educational context, the present chapter aimed to characterise the main theoretical approaches of studies that investigate science teaching mediated by ICT. We emphasise that the theoretical approaches of any study identify, characterise and list a set of studies and theories on a given subject. In this chapter, we will not seek to identify the theories but only the main characteristics of the theoretical approaches. We organised our reflection around two issues:

1. What are the main trends in the theoretical approaches for research on science teaching mediated by ICT?
2. How are the trends in theoretical approaches characterised in the context of science teaching mediated by ICT?

To better understand of theoretical approaches, we performed a literature search on different approaches to the learning and teaching of science mediated by ICT within formal education.

We also summarise current knowledge about the theoretical trends in research on ICT and the teaching of science, identify limitations of existing studies and provide guidelines for future research.

G. W. Rocha Fernandes et al., *Using ICT in Inquiry-Based Science Education*, SpringerBriefs in Education, https://doi.org/10.1007/978-3-030-17895-6_2

2.2   What Are the Main Trends in Theoretical Approaches for Science Education Mediated by ICT?

Based on the research in science education mediated by ICT, we organised the discussions about "theoretical approaches" of this chapter into five main trends:

(1) *Approaches to teaching and learning through the use of ICT*: show the influence of ICT on the teaching and learning of science.
(2) *Cognitive approaches*: present theories of teaching and learning, points of view related to cognition articulated with the use of ICT for learning science.
(3) *Approaches based on research, projects and case studies*: present curricular perspectives for the teaching and learning of science using ICT.
(4) *Approaches that emphasise conceptual knowledge*: present a discussion regarding "conceptual change", "conceptual evolution" and "conceptual understanding" of the scientific content provided by the use of ICT.

2.3   How Are the Trends in Theoretical Approaches Characterised in the Context of Science Education Mediated by ICT?

Science education involves an organised body of knowledge that is often abstract for many students (Bell & Bell, 2003; Scalise et al., 2011; She et al., 2012). To assist in the teaching and learning of concepts and scientific phenomena, the use of several media-driven technologies is highlighted within the classroom. Several studies seek answers regarding the real contributions of ICT in the school context and are based on the results of other studies or use different theories of teaching and learning to support their results. The following topics will seek to deepen the discussion on the main "theoretical approaches" in the context of science education mediated by ICT.

2.3.1   Approaches to Teaching and Learning Through the Use of ICT

The theoretical references (or theoretical approaches) evidenced in this topic are not related to a theoretical approach or theory but discuss the role of ICT in education, especially in the teaching and learning of scientific concepts and phenomena, i.e. show the influence of ICT on the teaching and learning of science.

The reflections in this first theoretical approach are based on 11 themes that classify the main ICT used to teaching and learning and its role in science education: simulation-based teaching and learning, Internet- and Web-based teaching and learning, computer-based teaching and learning, multimedia-based teaching and learning, technology-based teaching and learning, animation-based teaching and learning, mobile learning environments (MLE), hypermedia-based teaching and learning, MUVE-based teaching and learning, game-based teaching and learning and remote laboratory-based teaching and learning.

The main ICT used and its role in science education resemble the "cognitive tools" of Jonassen (2000) because the research does not provide a theory in its theoretical approach but rather tools for extension and cognitive restructuring for science learning. Jonassen (2000) conceptualises "cognitive tools" as "computer tools adapted or developed to function as students intellectual partners, in order to encourage and facilitate critical thinking and the learning of higher order (p. 21)".

Through these tools, various studies deepen the work of Mayer (2009) regarding *multimedia learning* bringing new perspectives on teaching and learning (Starbek, Starčič Erjavec, & Peklaj, 2010; Tolentino et al., 2009; Zheng, Yang, Garcia, & McCadden, 2008) in order to overcome the traditional teaching methods that still exist in many schools today.

We have to carefully analyse the theoretical assumptions in the main ICT used and its role in science education because there is also some tendency towards a "theory of technological education" (Bertrand, 2001) that "consists of a logical order of 'concrete' means to organize the teaching, regardless the nature of the content!" (p. 98). According to Bertrand (2001), a "theory of technological education" is detected mainly when a study (1) includes words related to process, communication, training, technology, techniques, computerised environments, interactive lab, hypermedia, individualised learning, etc.; (2) uses caution in talking about training or even instruction, instead of education; and (3) involves the use of communication technologies such as audio-visual equipment, DVDs, computers, etc.

When we look at the main ICT used in science education, we see that the discussions of theoretical approaches are in some instances very close to the "cognitive tools" of Jonassen (2000) and in others to the "theory of technological education" of Bertrand (2001). For example, in research on *simulations*, there are detailed studies about, among other things, the perception and spatial learning of students through simulations of scientific phenomena (Lindgren & Schwartz, 2009), the presence of necessary heuristics for the students to develop activities based on simulations (Veermans, van Joolingen, & de Jong, 2006), and even the use of simulations to relate theoretical studies to the real world using variable manipulation (Khan, 2010; Scalise et al., 2011).

The *Internet* and *Web* are also tools that appear in the theoretical approaches of various studies (Chin-Chung, 2009; Dimopoulos & Asimakopoulos, 2009; Gelbart, Brill, & Yarden, 2009; Katz, 2011; Kubasko, Jones, Tretter, & Andre, 2008; Lee et al., 2011; McCrory Wallace, Kupperman, Krajcik, & Soloway, 2000; She et al., 2012). We highlight the work of Lee et al. (2011), who conducted a major review of journal articles on Internet-based science learning from 1995 to 2008. This review is organised into two sections: (1) the role of demographics and characteristics of the students in Internet-based science learning and (2) the results of Internet-based science learning, such as attitude, motivation, conceptual understanding and conceptual change. Some important conclusions are drawn from this review. For example, the opportunity for students to have some control is essential for improving their attitudes and motivation for Internet-based science learning. However, the proper orientation of teachers and moderators is still crucial for Internet-based science learning.

The studies about "computer-based teaching and learning" or "computer-assisted instruction (CAI)" also exhibit some trends, in this case focused on the integration of ICT, with emphasis on the use of computers in the teaching of science. Several studies have sought to discuss the potential of the computer in the teaching of science (Clark & Jorde, 2004; Devolder et al., 2012; Pol, Harskamp, & Suhre, 2005; Ronen & Eliahu, 2000; Russell, Lucas, & McRobbie, 2004), mainly related to the development of experiments in virtual laboratories, problem-solving and the manipulation of variables.

The discussion of *multimedia-based teaching and learning* reinforces the principles of the *cognitive theory of multimedia learning* (Mayer, 2009) in that "people learn best through words and images than just through words" (Starbek et al., 2010; Tolentino et al., 2009; Zheng et al., 2008).

The studies that bring into their theoretical approaches "technology-based teaching and learning" discuss mainly the role of different technologies in the science curriculum, as well as the practices of the teachers in these scenarios (Hsu et al., 2008; Mayer-Smith, Pedretti, & Woodrow, 2000; Pedretti, Mayer-Smith, & Woodrow, 1998;Waight & Abd-El-Khalick, 2012).

The replacement of static images for animated images is also discussed on "animation-based teaching and learning". Science researchers and educators that address this issue believe that the use of animations to represent scientific models has a great potential to support the teaching and learning of scientific concepts (Barak, Ashkar, & Dori, 2011; Dalacosta, Kamariotaki-Paparrigopoulou, Palyvos, & Spyrellis, 2009).

Mobile learning environments (MLE) (Looi et al., 2011; Zhang et al., 2010) have been gaining attention in recent years, especially with the advent of *smartphones* and *tablets*.

Hypermedia-based teaching and learning is one of the main trends in the "theory of technological education" (Bertrand, 2001), examining technological environments from the perspective of interactivity and aiming to design increasingly interactive systems. For us, there is a pragmatic tendency, that is, more emphasis on systems that work than on a theoretical approach (Liu & Hmelo-Silver, 2009).

Currently, for scientific education, there are technology trends related to *multiuser virtual environments* (MUVE) or virtual worlds (Nelson, 2007), *game-based teaching and learning* (Squire & Jan, 2007) and *remote laboratory-based teaching and learning* (Lowe, Newcombe, & Stumpers, 2012).

The use of technological resources in the teaching of science has been addressed without considering the possible impact of teacher support, classroom context or the role of technological resources in the curriculum. We emphasise that although there are studies presenting theoretical approaches to technological descriptions, few contain deeper discussions of cognitive theories of knowledge, as discussed below.

2.3.2   Cognitive Approaches

This second topic brings assumptions related to cognitive and socio-cognitive theories to the use of ICT in science education. The reflections of the theoretical approaches of this topic emphasise actions and what is happening in the mind of the student when using an "educational technology" in relation to the understanding of the phenomena and of scientific concepts, i.e. present theories of teaching and learning, points of view related to cognition articulated with the use of ICT for learning science, for example:

(a) Social constructivism and sociocultural theory
(b) Constructivist approaches
(c) The effects of collaborative work
(d) Models and modelling to support science learning
(e) Dual-Situated Learning Model (DSLM)
(f) Dual coding theory
(g) Situated learning components
(h) Cognitive flexibility theory (CFT)
(i) Bruner's theory of intellectual development
(j) Learning styles
(k) Problem-solving
(l) Hierarchical structures in learning
(m) Student engagement in classrooms

The cognitive approaches present theoretical trends within cognitivist perspectives for the use of ICT in science education, with the three most common trends focusing on (a) *social constructivism and sociocultural theory*, (b) *constructivist approaches*, and (c) *the effects of collaborative work*.

For us, social constructivism has different characteristics from the original design of constructivism. From the constructivist point of view, learning occurs inside a person's mind. The students "learn by doing", accommodating new knowledge through experience and assimilating newly acquired knowledge in their current conceptual understanding (Yehudit Judy Dori & Belcher, 2005). From the social constructivist point of view, the construction of knowledge happens in a group environment, where the knowledge is distributed and shared; the role of social interaction is assumed to be a central element in the teaching and learning of science and the study of the world (Dori & Belcher, 2005). Students in pairs or in small groups help each other and all benefit from this interaction through the integration of knowledge from their peers and from the environment (Dori & Belcher, 2005).

The research of Cher Ping (2008) (*multi-user virtual environment*); Dori and Belcher (2005) (*simulation*); Furman and Barton (2006) (*video*); Kubasko et al. (2008) (*Internet and Web*); Lim, Nonis, and Hedberg (2006) (*game*); Oshima et al. (2004) (*Internet*); and Tolentino et al. (2009) (*virtual lab*) use an "educational technology" within the perspective of "social constructivism and sociocultural theory" to promote the teaching and learning of science. For example, we highlight here the study of Furberg and Ludvigsen (2008), who present the role of scientific argument as an aspect of social constructivism. Already, the studies of Byrne and Grace (2010) (*photographs and conceptual maps*), Jang (2006) and Ng and Gunstone (2002) (*Web*), Ronen and Eliahu (2000) (*simulation*) and Tekos and Solomonidou (2009) (*software*) present more constructivist perspectives in their theoretical approaches for science education.

In this sense, it is important to bring the discussion of "multimedia learning" proposed by Mayer and his collaborators (Mayer, 2009) together with the "dual coding theory" (DCT) (Paivio, 1990). The DCT suggests that the information submitted through verbal and visual channels are processed separately.

Dual processing presupposes that information is much easier to retain and retrieve when dual coded, because two independent codes are established in long-term memory (imagery and semantic). If these memory codes are linked, this increases the probability of recall and retrieval (Ardac & Akaygun, 2004, p. 319).

The DCT was tested in several multimedia studies to explain how the information presented through texts and animations produces learning (Mayer, 2009).

It is important to highlight that the DCT is different from the "Dual-Situated Learning Model (DSLM)", which is present, for example, in She and Lee (2008) (*web platform*), She and Liao (2010) (*hypermedia*) and Tseng, Tuan, and Chin (2010) (*animation*) and which is related to the use of various educational technologies as mechanisms of conceptual change regarding scientific phenomena and concepts. According to She and Liao (2010), "the DSLM emphasizes the nature of science concepts and students' ontological together with epistemological beliefs of science concepts, as its major theoretical constructs for conceptual change (p. 94)". The DSLM is related to studies on "conceptual change", which were very widespread in the 1990s (Mortimer, 1995). Situated learning indicates that for the conceptual change learning process, the nature of scientific concepts and students' beliefs about these science concepts should be considered to determine the essential mental sets[1] needed to construct a more scientific view of the concepts (She & Liao, 2010).

Many scientific concepts are difficult to understand (e.g. the density of the body, force, mass, etc.) and require more than a mental set, in particular for the construction of the concept. The term "dual" indicates that two essential components are important in order for conceptual change to occur and these components are interacting with one another. According to She and Liao (2010), there are three paired activities that are fundamental for the construction of each event in "dual situated learning". The first pair involves consideration for both the characteristics and the beliefs of the students in relation to scientific concepts. The second pair involves creating dissonance with the pre-existing knowledge of the students and providing a new mental set so that they can achieve a more scientific vision of the concept. The third pair involves awakening the students' motivation and challenging their ontological and epistemological beliefs regarding scientific concepts.

---

[1] A mental set includes an existing model for the representation of a particular phenomenon or information (She & Liao, 2010, p. 94).

The study by Hsu (2006) presents a theoretical discussion to use the ICT in science education on "situated learning components" (SLC). We note that the SLC differ from the "dual coding theory" (DCT) and the "Dual-Situated Learning Model (DSLM)". Hsu (2006) uses the concept of "situated learning" to emphasise that the understanding of scientific concepts occurs in real situations, in which knowledge is constructed through continuous interaction between human beings and situations. Thus, Hsu (2006) uses simulations with multiple representations (MR) that when "embedded in a 'situated learning' environment can provide the scaffolding and context in which students may explore the phenomena" (p. 541).

The relationship of the "models and modelling to support science learning" with the use of educational technologies appears in the studies of Ergazaki, Zogza, and Komis (2007) (*computer*), Pata and Sarapuu (2006) (*chat*), and Snir, Smith, and Raz (2003) (*software*). The educational value of modelling, which is a central activity in the scientific process, has a close relationship with the theoretical framework of *situated learning* in Hsu (2006) and with Johnson-Laird's study, which frames learning as a process of constructing mental models (Ergazaki et al., 2007).

Zydney and Grincewicz (2011) bring into their theoretical approaches the concept of "case-based issues" and base their study on the "cognitive flexibility theory (CFT)", in which a problem can be presented from multiple perspectives (Spiro, Collins, Thota, & Feltovich, 2003). "Computer learning environments based on Cognitive Flexibility Theory (CFT) are designed to prepare people to make adaptive responses to complex, novel, and dynamically evolving situations" (Spiro et al., 2003, p. 8). For Zydney and Grincewicz (2011), this theory helps in the design of case-based issues within technology-based environments, providing more specific design principles.

Other cognitive approaches, less mentioned in the literature to the use of ICT in science education, but as important as the others, include *Bruner's theory of intellectual development* (Mayo, Sharma, & Muller, 2009), *learning styles* (Sun, Lin, & Yu, 2008), *problem-solving* (Shin, Jonassen, & McGee, 2003), *hierarchical structures in learning* (Shapiro, 1999) and *student engagement in classrooms* (Wu & Huang, 2007).

2.3.3 Approaches Based on Research, Projects and Case Studies

The development of computers and various tools associated with them provides educators with several possibilities to stimulate students for independent study. There is a set of possibilities for the use of educational technologies to stimulate students' abilities to investigate, discover, observe, solve problems, develop projects, perform activities and deepen knowledge about scientific cases. These possibilities exemplify contemporary perspectives related to the *development of new curricula*, which in recent decades has been gaining attention in relation to science education. The main themes of ICT use in the curriculum from a contemporary perspective are:

(a) Inquiry-based approach
(b) Discovery-based approach
(c) Case study-based approach
(d) Problem-based approach
(e) Project-based approach
(f) Observation-based approach
(g) Activity-based approach

These themes appear in theoretical approaches regarding the use of various educational technologies in science education and deserve to be analysed individually.

2.3.3.1   The Inquiry-Based Approach

Currently, investigative activities do not resemble those that imitate the "scientific method", mainly characterised by international projects on science education in the United States and Europe from the 1960s. Learning by inquiry, within the perspective of social constructivism, involves the active involvement of students in the exploration of some phenomena of daily life in a way that leads them to ask questions, generate hypotheses, share ideas and build knowledge (Driver, Hilary, John, Mortimer, & Philip, 1999; Driver, Newton, & Osborne, 2000). To include research as an activity for scientific education, several technologies have been used in the science curriculum (Ebenezer, Kaya, & Ebenezer, 2011). For example, Jaakkola and Nurmi (2008) showed that the combination of virtual simulation with laboratory resources leads to learning gains that are statistically greater than the use of any simulation or laboratory activities alone and also promotes more efficiently the students' conceptual understanding. Mistler-Jackson and Butler Songer (2000) observed how different sixth-grade students learned content related to the atmosphere using technology, before and after the implementation of a program that used data available on the Internet. Hakkarainen (2003), using a collaborative learning perspective, examined the research processes of 28 students from fifth and sixth grade in computer-supported learning. The results of the study indicated that, with the guidance of the teacher, the students were able to produce intuitive explanations of biological phenomena, guide their own learning processes, continue with their own research topics and be engaged in constructive interaction between pairs that helped them to go beyond their intuitive explanations and towards theoretical, scientific explanations.

The results of inquiry-based approach are positive, and several other studies are promising for a research-based understanding of science education mediated by various educational technologies (Barab, Sadler, Heiselt, Hickey, & Zuiker, 2007; Dori & Sasson, 2008; Ebenezer et al., 2011; Lin, Hsu, & Yeh, 2012; Lin, Wang, & Lin, 2012; Pyatt & Sims, 2012).

2.3.3.2   The Discovery-Based Approach

This curriculum perspective, sometimes regarded as a synonym of the "inquiry-based approach" (Jaakkola & Nurmi, 2008), has also been gaining prominence in various studies. In order to understand the role of educational technologies in learning-by-discovery, several significant studies have been conducted regarding science learning (Reid, Zhang, & Chen, 2003; Zhang, Chen, Sun, & Reid, 2004). For example, according to Reid et al. (2003) and Zhang et al. (2004), the learning of science mediated by an environment with simulations must be grounded in a learning-by-discovery framework that is organised around three perspectives: *interpretative support* (S), *experimental support* (ES) and *reflective support* (RS). According to these authors, these three perspectives make learning-by-discovery significant, systematic and reflexive.

2.3.3.3   Case Study-Based Approach

"Case studies", also known as "case narratives", "stories with a message" or "stories to educate" (Dori & Sasson, 2008), have been gaining attention in the teaching and research of science. Originating from the faculties of administration and medicine, the "case study" method is usually based on real-life stories (though the stories may be fictitious), which serve as examples to analyse and apply and which make science more relevant to the daily lives of students (Dori & Sasson, 2008). The cases may be open or closed and may not require one single correct answer but rather several possible solutions to a dilemma. The cases may contain scientific aspects that involve emotions, ethics or politics, may present unresolved dilemmas or socio-scientific issues and may produce multiple solutions. For example, the study by Zydney and Grincewicz (2011) sought to understand the position of the students before a case mediated by videos (within a multimedia learning environment). The case involved a complex problem for the students from a socio-scientific point of view. A qualitative analysis of the students' reflections indicated that many of them appreciated the complexity, authenticity and ethical dimensions of the problem. This study has provided some preliminary evidence that offering students the opportunity to watch videos from different perspectives can encourage them to think in alternative ways about a complex problem.

2.3.3.4   The Problem-Based Approach

The PBL (problem-based learning) method is a pedagogical/didactic strategy centred on the student. This approach is very close to the "inquiry-based approach" and the "discovery-based approach". In this method, the student is no longer a passive element, exposed to information through classes; rather, the student seeks knowledge for problem-solving. In a more contemporary perspective, different from that of the 1920s when the approach emerged in the United States, the "problem-based approach" has been used in several techniques for developing the scientific knowledge of students. One of these possibilities is the use of some educational technologies. For example, the research of Barab et al. (2009) used video games with 51 graduate students to support their study regarding the use of pesticides in agriculture and the production of waste by society. The research by Ketelhut (2007) used a MUVE called *River City*, which is student-centred and based on issues for which students can gather evidence from the environment and can experience, in part, the reality of an epidemiologist investigating the outbreak of a disease.

2.3.3.5   The Project-Based Approach

The "project-based approach" or "project-based learning" has as its main theoretical principles "Educating by Research" and "Pedagogy of Projects" and dates from the beginning of the twentieth century, having supporters such as John Dewey and William Kilpatrick. The development of projects through research in the classroom, besides helping the students in the construction of meaningful knowledge, enables them to intervene in the situations in which they live. The approach is interested in overcoming the fragmentation of content and tends towards interdisciplinary and transdisciplinary perspectives for the contextualization of the curriculum. The use of educational technologies can support the process of research in the constructivist perspective. The incorporation of technology and the constructivist approach can positively influence the understanding and learning of scientific concepts (Wang, Ke, Wu, & Hsu, 2012). In this perspective, we have two examples. The first is the study by Hansson, Redfors, and Rosberg (2011), in which the researchers and teachers together design and implement a digital learning environment to describe the results of the students' decisions in relation to different socio-scientific issues and the types of support that they use for their decisions. The second is the study by Wang et al. (2012), in which sixth-grade students use blogs, Microsoft PowerPoint (PPT) and the Internet as tools for learning diverse science content based on projects and collaborative work.

2.3.3.6   The Observation-Based Approach

This approach has a very close relationship with the "inquiry-based approach" and the "discovery-based approach". The observation of natural phenomena allows the student to relate scientific theories with the real world, to become familiar with the use of equipment and relevant techniques of scientific research through a set of appropriate actions (Kong, Yeung, & Wu, 2009). In this perspective, we found the study by Kong et al. (2009), who explored the result of students' learning after the collection of data by a remotely controlled system called LabVNC, which is free software. They also explored the views of a teacher on the use of LabVNC in science education.

2.3.3.7   The Activity-Based Approach

This approach is a form of work/activity including teams, groups, organizations, etc. beyond just an actor or user. Dori and Belcher (2005) use this approach to analyse the cognitive and affective effects of a special space for learning, rich in media, called project *Technology-Enabled Active Learning* (TEAL) at the Massachusetts Institute of Technology (MIT). This approach, which characterises the study of Dori and Belcher (2005), includes the objective, content, artefacts, rules, teams and division of labour among students in relation to the media-rich space.

In short, we see that in these approaches there are contemporary trends in the integration of educational technology in science education with prospects for transformation of the science curricula, but they seem to deserve fuller attention from teachers and researchers.

2.3.4   Approaches that Emphasise Conceptual Knowledge

This theoretical approach features the role of ICT in promoting "conceptual change" or the "conceptual understanding" of science. It features an apparent relationship between conceptual change and conceptual understanding, but the characteristics of this relationship remain uncertain.

The theoretical approaches that emphasise conceptual knowledge can be thought of as follows:

(a) The role of ICT in promoting conceptual change in science
(b) The role of ICT in promoting conceptual understanding in science

The first theoretical approach, *the role of ICT* in promoting conceptual change in science, contains studies that bring into their theoretical frameworks the classical view of conceptual change by Posner, Strike, Hewson, and Gertzog (1982). In other words, these studies see conceptual change as the replacement of one design with another in the cognitive structure of the learner, thus promoting conceptual understandings that more closely resemble accepted scientific concepts. Several studies mediated by educational technologies have supported this theory. For example, Jaakkola and Nurmi (2008) investigated the best "association" for teaching the concepts of electricity to fourth- and fifth-grade students, given that this concept contains several alternative and intuitive concepts. Thus, the authors used virtual simulations and real objects to demonstrate the concepts of electricity. The combination of virtual simulations with lab resources led to greater learning gains compared with other simulations or laboratory activity alone and more efficiently promoted the students' conceptual understanding. Although the authors utilise "inquiry-based (or discovery-based) learning" as support for conceptual change, their conceptual framework reflects the classical theory and does not consider that the intuitive conceptions of students may return in other contexts related to the study of electricity.

The study by Jacobson and Archodidou (2000) proposes the development of a hypermedia tool (*knowledge mediator framework – KMF*) developed in the model of *problem-based learning* in order to achieve significant gains in learning, such as in conceptual understanding, conceptual change and the transfer of knowledge. The study features thoughts and definitions related to *conceptual structures, conceptual visualizations, conceptual scaffolding* and *conceptual crisscrossing*. Following is the study by Vosniadou and Brewer (1994) that uses the referential of *mental models* to explain student difficulties in the understanding of scientific concepts but does not clearly overcome the traditional perspective regarding conceptual change.

The researchers Li, Law, and Lui (2006) examined how the conceptual change of 20 sixth-grade students progresses during the process of constructing explanatory models for understanding the phenomenon of evaporation. The authors used modelling software called *World Maker 2000* in conjunction with a strategy of cognitive disturbance performed by a professor. They found that the use of the software and the teacher strategy were effective in helping students to trade their alternative concepts for correct scientific concepts, but the researchers did not deepen the paths of conceptual change between the groups.

The study by Park, Khan, and Petrina (2009) also presents the classical vision of "conceptual change". Park et al. (2009) analysed the contributions of "computer-assisted instruction (CAI)" in science classes with 234 Korean middle-school students. This previous study cites Posner et al. (1982), but the authors do not discuss current theories accepted in the "conceptual change" model.

On the other hand, there are theoretical approaches of "conceptual change" not as a replacement of one design with another but as a "dual" process (She & Lee, 2008; She & Liao, 2010; Tseng et al., 2010). Over the past 30 years, the perspectives on conceptual change held by researchers in science education (e.g. Posner et al., 1982) have differed substantially from those of cognitive psychologists (Vosniadou & Brewer, 1994). To reconcile these differences, some studies (She & Lee, 2008; She & Liao, 2010) have integrated the strengths of both sides in a theoretical construction for the development of *Dual-Situated Learning Model (DSLM)* (see "Cognitive Approaches").

Another theoretical approach refers to "the role of ICT in promoting conceptual understanding in science" (Chang, Quintana, & Krajcik, 2010; Chang, Yeh, & Barufaldi, 2010; Hoffman, Wu, Krajcik, & Soloway, 2003; Olympiou & Zacharia, 2012; Zacharia, Olympiou, & Papaevripidou, 2008). Hoffman et al. (2003) develop the concept and the theoretical bases regarding "content understanding". According to these authors, "content understanding" is not only a remembrance of facts and definitions associated with a particular subject area but also the use of mapping schemes to associate concepts with references and memory strategies:

> Similarly, content understanding can be viewed as a matter of degree in which an individual understands concepts, principles, structures, or processes at a relatively deep level and is able to demonstrate certain behaviours. (Hoffman et al., 2003, p. 324)

In this perspective, Hoffman et al. (2003) investigated the nature of the understanding of scientific content, the use of the research and the strategies used by 830 sixth-grade students for using the Internet (via Artemis, an interface in the Digital Library) to conduct searches about astronomy, ecology, geology or weather. The analysis of the data showed that the students had built significant understandings through online research, although the accuracy and depth of their understandings varied.

Olympiou and Zacharia (2012) and Zacharia et al. (2008) compared the effectiveness, with respect to changes in students' understanding of scientific concepts, of experiences with materials or *physical manipulatives* (*PM*) alone and experiences with PM and *virtual manipulatives* (*VM*), with the use of PM preceding the use of VM. The results indicated that experiences with the combination of PM and VM improve the conceptual understanding of students more than experiences with the PM alone. It is important to highlight that the use of VM was identified as the cause of differentiation; however, not all studies present the same results when using a combination of real and virtual experiments.

Chang, Quintana and Krajcik (2010) explored the use of different activities mediated by an animation tool to promote the understanding of chemistry for high school students. However, other authors are aware that animations alone may not be sufficient to improve student understanding (Barak et al., 2011; Dalacosta et al., 2009) and different teaching methods that use animation to promote the understanding of scientific content must be considered (Mayer, 2009).

Despite the recurring criticisms of the model of conceptual change, studies continue to be conducted within this construct, though mediated by educational technologies. Although numerous studies show that there is no conceptual change in the mind of the student, the expression is already instilled in the literature and its use is widespread (Mortimer, 1995). Many more people are possibly thinking along the same lines, which is probably a sign that it is time to finally abandon the term "conceptual change" and models that suggest "conceptual change". The terms *evolution, development, enrichment* and *conceptual discrimination of meanings* are the most promising ideas because they do not involve the change of concepts or meanings (Moreira & Greca, 2003).

2.4 Conclusions

When we did the preparation of this chapter and presented information about the use of educational technologies in science education, we proposed to understand the characteristics of the main theoretical approaches to the use of ICT in science education. Based on our reading and analysis, we proposed that the main theoretical trends are organised into four major topics of interest to researchers and science teachers: (1) approaches to teaching and learning through the use of ICT; (2) cognitive approaches; (3) approaches based on inquiry, research, projects and case studies; and (4) approaches that emphasise conceptual knowledge.

We could verify that these theoretical trends are characterised in different ways in the context of science education mediated by ICTs. Theoretical assumptions tend to approach a "technological theory of education" because many references are concerned with the process, communication, training, technology, techniques, computerised environments, interactive laboratories, hypermedia, individualised and mobile teaching, the Internet, games, etc.

For us, the use of technologies can be supported on the basis of social constructivism because working with educational technologies happens in a group atmosphere, where knowledge is distributed and shared. Students help each other in pairs, and the construction of knowledge occurs through the integration of knowledge from their peers and the environment (Dori & Belcher, 2005).

There is a strong tendency towards the use of educational technologies, particularly the computer, to stimulate students' abilities to investigate, discover, observe, solve problems, develop projects, carry out activities and deepen scientific understandings. With all these possibilities, there are theoretical approaches that seek to explore "conceptual understanding" mediated by various educational technologies that do not deepen the concept of conceptual change, except for those studies that use the "Dual-Situated Learning Model (DSLM)".

Finally, the theoretical approaches identified in this study are not far beyond those already discussed in the literature. There is a lack of broader discussion on the theoretical foundations that support the use of ICT in science education, and those theoretical approaches are different from the main perspectives identified here. The main contribution of the highlighted data is in the understanding that the use of ICT in science education is not an isolated action without a theoretical basis. The use of ICT is still planned and supported by traditional theoretical trends in teaching, learning, knowledge and curriculum proposals.

Of course, like any other educational tool, the effectiveness of ICT is limited by the ways in which the technologies are used, and indeed, the theoretical approaches used to support the use of ICT in science education should be considered.

References

Ardac, D., & Akaygun, S. (2004). Effectiveness of multimedia-based instruction that emphasizes molecular representations on students' understanding of chemical change. *Journal of Research in Science Teaching, 41*, 317–337. https://doi.org/10.1002/tea.20005

Barab, S., Sadler, T., Heiselt, C., Hickey, D., & Zuiker, S. (2007). Relating narrative, inquiry, and inscriptions: Supporting consequential play. *Journal of Science Education and Technology, 16*, 59–82. https://doi.org/10.1007/s10956-006-9033-3

Barab, S., Scott, B., Siyahhan, S., Goldstone, R., Ingram-Goble, A., Zuiker, S., & Warren, S. (2009). Transformational play as a curricular scaffold: Using videogames to support science education. *Journal of Science Education and Technology, 18*, 305–320. https://doi.org/10.1007/s10956-009-9171-5

Barak, M., Ashkar, T., & Dori, Y. J. (2011). Learning science via animated movies: Its effect on students' thinking and motivation. *Computers & Education, 56*, 839–846. https://doi.org/10.1016/j.compedu.2010.10.025

Bell, R., & Bell, L. (2003). A bibliography of articles on instructional technology in science educa-tion. *Contemporary Issues in Technology and Teacher Education, 2*(4). Retrieved from http://www.citejournal.org/vol2/iss4/science/article2.cfm

Bertrand, Y. (2001). *Teorias contemporâneas da educação* (2nd ed.). Lisboa, Portugal: Editora Piaget.

Byrne, J., & Grace, M. (2010). Using a concept mapping tool with a photograph association tech-nique (compat) to elicit children's ideas about microbial activity. *International Journal of Science Education, 32*, 479–500. https://doi.org/10.1080/09500690802688071

Chang, C., Yeh, T., & Barufaldi, J. P. (2010). The positive and negative effects of science concept tests on student conceptual understanding. *International Journal of Science Education, 32*, 265–282. https://doi.org/10.1080/09500690802650055

Chang, H.-Y., Quintana, C., & Krajcik, J. S. (2010). The impact of designing and evaluating molecular animations on how well middle school students understand the particulate nature of matter. *Science Education, 94*, 73–94. https://doi.org/10.1002/sce.20352

Cher Ping, L. (2008). Global citizenship education, school curriculum and games: Learning math-ematics, English and science as a global citizen. *Computers & Education, 51*, 1073–1093. https://doi.org/10.1016/j.compedu.2007.10.005

Chin-Chung, T. (2009). Conceptions of learning versus conceptions of web-based learning: The differences revealed by college students. *Computers & Education, 53*(4), 1092–1103. https://doi.org/10.1016/j.compedu.2009.05.019

Clark, D., & Jorde, D. (2004). Helping students revise disruptive experientially supported ideas about thermodynamics: Computer visualizations and tactile models. *Journal of Research in Science Teaching, 41*, 1–23. https://doi.org/10.1002/tea.10097

Dalacosta, K., Kamariotaki-Paparrigopoulou, M., Palyvos, J. A., & Spyrellis, N. (2009). Multimedia application with animated cartoons for teaching science in elementary education. *Computers & Education, 52*, 741–748. https://doi.org/10.1016/j.compedu.2008.11.018

Devolder, A., van Braak, J., & Tondeur, J. (2012). Supporting self-regulated learning in computer-based learning environments: Systematic review of effects of scaffolding in the domain of science education. *Journal of Computer Assisted Learning, 28*, 557–573. https://doi.org/10.1111/j.1365-2729.2011.00476.x

Dimopoulos, K., & Asimakopoulos, A. (2009). Science on the web: Secondary school students' navigation patterns and preferred pages' characteristics. *Journal of Science Education and Technology, 19*, 246–265. https://doi.org/10.1007/s10956-009-9197-8

Dori, Y. J., & Belcher, J. (2005). How does technology-enabled active learning affect undergradu-ate students' understanding of electromagnetism concepts? *Journal of the Learning Sciences, 14*, 243–279. https://doi.org/10.1207/s15327809jls1402_3

Dori, Y. J., & Sasson, I. (2008). Chemical understanding and graphing skills in an honors case-based computerized chemistry laboratory environment: The value of bidirectional visual and textual representations. *Journal of Research in Science Teaching, 45*, 219–250. https://doi.org/10.1002/tea.20197

Driver, R., Hilary, A., John, L., Mortimer, E. F., & Philip, S. (1999). Construindo conhecimento científico na sala de aula. *Química Nova na Escola, 1*(9).

Driver, R., Newton, P., & Osborne, J. (2000). Establishing the norms of scientific argumenta-tion in classrooms. *Science Education, 84*(3), 287–312. https://doi.org/10.1002/(SICI)1098-237X(200005)84:3<287::AID-SCE1>3.0.CO;2-A

Ebenezer, J., Kaya, O. N., & Ebenezer, D. L. (2011). Engaging students in environmental research projects: Perceptions of fluency with innovative technologies and levels of scientific inquiry abilities. *Journal of Research in Science Teaching, 48*, 94–116. https://doi.org/10.1002/tea.20387

Ergazaki, M., Zogza, V., & Komis, V. (2007). Analysing students' shared activity while modeling a biological process in a computer-supported educational environment. *Journal of Computer Assisted Learning, 23*, 158–168. https://doi.org/10.1111/j.1365-2729.2006.00214.x

Furberg, A., & Ludvigsen, S. (2008). Students' meaning-making of socio-scientific issues in computer mediated settings: Exploring learning through interaction trajectories. *International Journal of Science Education, 30*, 1775–1799. https://doi.org/10.1080/09500690701543617

Furman, M., & Barton, A. C. (2006). Capturing urban student voices in the creation of a science mini-documentary. *Journal of Research in Science Teaching, 43*, 667–694. https://doi.org/10.1002/tea.20164

Gelbart, H., Brill, G., & Yarden, A. (2009). The impact of a web-based research simulation in bioinformatics on students' understanding of genetics. *Research in Science Education, 39*, 725–751. https://doi.org/10.1007/s11165-008-9101-1

Hakkarainen, K. (2003). Progressive inquiry in a computer-supported biology class. *Journal of Research in Science Teaching, 40*, 1072–1088. https://doi.org/10.1002/tea.10121

Hansson, L., Redfors, A., & Rosberg, M. (2011). Students' socio-scientific reasoning in an astrobiological context during work with a digital learning environment. *Journal of Science Education and Technology, 20*, 388–402. https://doi.org/10.1007/s10956-010-9260-5

Hoffman, J. L., Wu, H.-K., Krajcik, J. S., & Soloway, E. (2003). The nature of middle school learners' science content understandings with the use of on-line resources. *Journal of Research in Science Teaching, 40*, 323–346. https://doi.org/10.1002/tea.10079

Hsu, Y.-S. (2006). Lesson rainbow: The use of multiple representations in an internet-based, discipline-integrated science lesson. *British Journal of Educational Technology, 37*, 539–557. https://doi.org/10.1111/j.1467-8535.2006.00551.x

Hsu, Y.-S., Wu, H.-K., & Hwang, F.-K. (2008). Fostering high school students' conceptual understandings about seasons: The design of a technology-enhanced learning environment. *Research in Science Education, 38*, 127–147. https://doi.org/10.1007/s11165-007-9041-1

Jaakkola, T., & Nurmi, S. (2008). Fostering elementary school students' understanding of simple electricity by combining simulation and laboratory activities. *Journal of Computer Assisted Learning, 24*, 271–283. https://doi.org/10.1111/j.1365-2729.2007.00259.x

Jacobson, M. J., & Archodidou, A. (2000). The design of hypermedia tools for learning: Fostering conceptual change and transfer of complex scientific knowledge. *Journal of the Learning Sciences, 9*, 145–199. https://doi.org/10.1207/s15327809jls0902_2

Jang, S. (2006). The effects of incorporating web-assisted learning with team teaching in seventh-grade science classes. *International Journal of Science Education, 28*, 615–632. https://doi.org/10.1080/09500690500339753

Jonassen, D. H. (2000). *Computadores, ferramentas cognitivas: Desenvolver o pensamento crítico nas escolas* (2nd ed.). Porto, Portugal: Porto Editora.

Katz, P. (2011). A case study of the use of internet photobook technology to enhance early childhood "scientist" identity. *Journal of Science Education and Technology, 20*, 525–536. https://doi.org/10.1007/s10956-011-9301-8

Ketelhut, D. (2007). The impact of student self-efficacy on scientific inquiry skills: An exploratory investigation in river city, a multi-user virtual environment. *Journal of Science Education and Technology, 16*, 99–111. https://doi.org/10.1007/s10956-006-9038-y

Khan, S. (2010). New pedagogies on teaching science with computer simulations. *Journal of Science Education and Technology, 20*(3), 215–232. https://doi.org/10.1007/s10956-010-9247-2

Kong, S. C., Yeung, Y. Y., & Wu, X. Q. (2009). An experience of teaching for learning by observation: Remote-controlled experiments on electrical circuits. *Computers & Education, 52*, 702–717. https://doi.org/10.1016/j.compedu.2008.11.011

Kubasko, D., Jones, M. G., Tretter, T., & Andre, T. (2008). Is it live or is it memorex? Students' synchronous and asynchronous communication with scientists. *International Journal of Science Education, 30*, 495–514. https://doi.org/10.1080/09500690701217220

Lee, S. W., Tsai, C., Wu, Y., Tsai, M., Liu, T., Hwang, F., et al. (2011). Internet-based science learning: A review of journal publications. *International Journal of Science Education, 33*, 1893–1925. https://doi.org/10.1080/09500693.2010.536998

Li, S. C., Law, N., & Lui, K. F. A. (2006). Cognitive perturbation through dynamic modelling: A pedagogical approach to conceptual change in science. *Journal of Computer Assisted Learning, 22*, 405–422. https://doi.org/10.1111/j.1365-2729.2006.00187.x

Lim, C. P., Nonis, D., & Hedberg, J. (2006). Gaming in a 3D multiuser virtual environment: Engaging students in science lessons. *British Journal of Educational Technology, 37*, 211–231. https://doi.org/10.1111/j.1467-8535.2006.00531.x

Lin, J. M., Wang, P., & Lin, I. (2012). Pedagogy technology: A two-dimensional model for teachers' ICT integration. *British Journal of Educational Technology, 43*(1), 97–108. https://doi.org/10.1111/j.1467-8535.2010.01159.x

Lin, L.-F., Hsu, Y.-S., & Yeh, Y.-F. (2012). The role of computer simulation in an inquiry-based learning environment: Reconstructing geological events as geologists. *Journal of Science Education and Technology, 21*, 370–383. https://doi.org/10.1007/s10956-011-9330-3

Lindgren, R., & Schwartz, D. L. (2009). Spatial learning and computer simulations in science. *International Journal of Science Education, 31*, 419–438. https://doi.org/10.1080/09500690802595813

Liu, L., & Hmelo-Silver, C. E. (2009). Promoting complex systems learning through the use of conceptual representations in hypermedia. *Journal of Research in Science Teaching, 46*, 1023–1040. https://doi.org/10.1002/tea.20297

Looi, C.-K., Zhang, B., Chen, W., Seow, P., Chia, G., Norris, C., & Soloway, E. (2011). 1:1 mobile inquiry learning experience for primary science students: A study of learning effectiveness. *Journal of Computer Assisted Learning, 27*, 269–287. https://doi.org/10.1111/j.1365-2729.2010.00390.x

Lowe, D., Newcombe, P., & Stumpers, B. (2012). Evaluation of the use of remote laboratories for secondary school science education. *Research in Science Education, 43*, 1–23. https://doi.org/10.1007/s11165-012-9304-3

Mayer, R. E. (2009). Teoria cognitiva da aprendizagem multimédia. In *Ensino online e aprendizagem multimédia*. Lisboa, Portugal: Relógio D'Água Editores.

Mayer-Smith, J., Pedretti, E., & Woodrow, J. (2000). Closing of the gender gap in technology enriched science education: A case study. *Computers & Education, 35*, 51–63. https://doi.org/10.1016/S0360-1315(00)00018-X

Mayo, A., Sharma, M., & Muller, D. (2009). Qualitative differences between learning environments using videos in small groups and whole class discussions: A preliminary study in physics. *Research in Science Education, 39*, 477–493. https://doi.org/10.1007/s11165-008-9090-0

McCrory Wallace, R., Kupperman, J., Krajcik, J., & Soloway, E. (2000). Science on the web: Students online in a sixth-grade classroom. *Journal of the Learning Sciences, 9*, 75–104. https://doi.org/10.1207/s15327809jls0901_5

Mistler-Jackson, M., & Butler Songer, N. (2000). Student motivation and internet technology: Are students empowered to learn science? *Journal of Research in Science Teaching, 37*, 459–479. https://doi.org/10.1002/(SICI)1098-2736(200005)37:5<459::AID-TEA5>3.0.CO;2-C

Moreira, M. A., & Greca, I. M. (2003). Conceptual change: Critical analysis and proposals in the light of the meaningful learning theory. *Ciência & Educação, 9*(2), 301–315. https://doi.org/10.1590/S1516-73132003000200010

Mortimer, E. F. (1995). Conceptual change or conceptual profile change? *Science & Education, 4*(3), 267–285. https://doi.org/10.1007/BF00486624

Nelson, B. (2007). Exploring the use of individualized, reflective guidance in an educational multiuser virtual environment. *Journal of Science Education and Technology, 16*, 83–97. https://doi.org/10.1007/s10956-006-9039-x

Ng, W., & Gunstone, R. (2002). Students' perceptions of the effectiveness of the world wide web as a research and teaching tool in science learning. *Research in Science Education, 32*, 489–510. https://doi.org/10.1023/A:1022429900836

Olympiou, G., & Zacharia, Z. C. (2012). Blending physical and virtual manipulatives: An effort to improve students' conceptual understanding through science laboratory experimentation. *Science Education, 96*, 21–47. https://doi.org/10.1002/sce.20463

Oshima, J., Oshima, R., Murayama, I., Inagaki, S., Takenaka, M., Nakayama, H., & Yamaguchi, E. (2004). Design experiments in Japanese elementary science education with computer support for collaborative learning: Hypothesis testing and collaborative construction. *International Journal of Science Education, 26*, 1199–1221. https://doi.org/10.1080/0950069032000138824

Paivio, A. (1990). *Mental representations: A dual coding approach*. New York, NY: Oxford University Press.

Park, H., Khan, S., & Petrina, S. (2009). ICT in science education: A quasi-experimental study of achievement, attitudes toward science, and career aspirations of Korean middle school students. *International Journal of Science Education, 31*, 993–1012. https://doi.org/10.1080/09500690701787891

Pata, K., & Sarapuu, T. (2006). A comparison of reasoning processes in a collaborative modelling environment: Learning about genetics problems using virtual chat. *International Journal of Science Education, 28*, 1347–1368. https://doi.org/10.1080/09500690500438670

Pedretti, E., Mayer-Smith, J., & Woodrow, J. (1998). Technology, text, and talk: Students' perspectives on teaching and learning in a technology-enhanced secondary science classroom. *Science Education, 82*, 569–589. https://doi.org/10.1002/(SICI)1098-237X(199809)82:5<569 ::AID-SCE3>3.0.CO;2-7

Pol, H., Harskamp, E., & Suhre, C. (2005). Solving physics problems with the help of computer-assisted instruction. *International Journal of Science Education, 27*, 451–469. https://doi.org/10.1080/0950069042000266164

Posner, G. J., Strike, K. A., Hewson, P. W., & Gertzog, W. A. (1982). Accommodation of a scientific conception: Toward a theory of conceptual change. *Science Education, 66*(2), 211–227. https://doi.org/10.1002/sce.3730660207

Pyatt, K., & Sims, R. (2012). Virtual and physical experimentation in inquiry-based science labs: Attitudes, performance and access. *Journal of Science Education and Technology, 21*, 133–147. https://doi.org/10.1007/s10956-011-9291-6

Reid, D. J., Zhang, J., & Chen, Q. (2003). Supporting scientific discovery learning in a simulation environment. *Journal of Computer Assisted Learning, 19*, 9–20. https://doi.org/10.1046/j.0266-4909.2003.00002.x

Ronen, M., & Eliahu, M. (2000). Simulation — a bridge between theory and reality: The case of electric circuits. *Journal of Computer Assisted Learning, 16*, 14–26. https://doi.org/10.1046/j.1365-2729.2000.00112.x

Russell, D. W., Lucas, K. B., & McRobbie, C. J. (2004). Role of the microcomputer-based laboratory display in supporting the construction of new understandings in thermal physics. *Journal of Research in Science Teaching, 41*, 165–185. https://doi.org/10.1002/tea.10129

Scalise, K., Timms, M., Moorjani, A., Clark, L., Holtermann, K., & Irvin, P. S. (2011). Student learning in science simulations: Design features that promote learning gains. *Journal of Research in Science Teaching, 48*, 1050–1078. https://doi.org/10.1002/tea.20437

Shapiro, A. M. (1999). The relevance of hierarchies to learning biology from hypertext. *Journal of the Learning Sciences, 8*, 215–243. https://doi.org/10.1207/s15327809jls0802_2

She, H.-C., Cheng, M.-T., Li, T.-W., Wang, C.-Y., Chiu, H.-T., Lee, P.-Z., et al. (2012). Web-based undergraduate chemistry problem-solving: The interplay of task performance, domain knowledge and web-searching strategies. *Computers & Education, 59*, 750–761. https://doi.org/10.1016/j.compedu.2012.02.005

She, H.-C., & Lee, C.-Q. (2008). SCCR digital learning system for scientific conceptual change and scientific reasoning. *Computers & Education, 51*, 724–742. https://doi.org/10.1016/j.compedu.2007.07.009

She, H.-C., & Liao, Y.-W. (2010). Bridging scientific reasoning and conceptual change through adaptive web-based learning. *Journal of Research in Science Teaching, 47*, 91–119. https://doi.org/10.1002/tea.20309

Shin, N., Jonassen, D. H., & McGee, S. (2003). Predictors of well-structured and ill-structured problem solving in an astronomy simulation. *Journal of Research in Science Teaching, 40*, 6–33. https://doi.org/10.1002/tea.10058

Smetana, L. K., & Bell, R. L. (2012). Computer simulations to support science instruction and learning: A critical review of the literature. *International Journal of Science Education, 34,* 1337–1370. https://doi.org/10.1080/09500693.2011.605182

Snir, J., Smith, C. L., & Raz, G. (2003). Linking phenomena with competing underlying models: A software tool for introducing students to the particulate model of matter. *Science Education, 87*(6), 794–830. https://doi.org/10.1002/sce.10069

Spiro, R. J., Collins, B. P., Thota, J. J., & Feltovich, P. J. (2003). Cognitive flexibility theory: Hypermedia for complex learning, adaptive knowledge application, and experience acceleration. *Educational Technology, 43*(5), 5–10.

Squire, K., & Jan, M. (2007). Mad city mystery: Developing scientific argumentation skills with a place-based augmented reality game on handheld computers. *Journal of Science Education and Technology, 16,* 5–29. https://doi.org/10.1007/s10956-006-9037-z

Starbek, P., Starčič Erjavec, M., & Peklaj, C. (2010). Teaching genetics with multimedia results in better acquisition of knowledge and improvement in comprehension. *Journal of Computer Assisted Learning, 26,* 214–224. https://doi.org/10.1111/j.1365-2729.2009.00344.x

Sun, K., Lin, Y., & Yu, C. (2008). A study on learning effect among different learning styles in a web-based lab of science for elementary school students. *Computers & Education, 50,* 1411–1422. https://doi.org/10.1016/j.compedu.2007.01.003

Tekos, G., & Solomonidou, C. (2009). Constructivist learning and teaching of optics concepts using ICT tools in Greek primary school: A pilot study. *Journal of Science Education and Technology, 18,* 415–428. https://doi.org/10.1007/s10956-009-9158-2

Tolentino, L., Birchfield, D., Megowan-Romanowicz, C., Johnson-Glenberg, M. C., Kelliher, A., & Martinez, C. (2009). Teaching and learning in the mixed-reality science classroom. *Journal of Science Education and Technology, 18,* 501–517. https://doi.org/10.1007/s10956-009-9166-2

Tseng, C., Tuan, H., & Chin, C. (2010). Investigating the influence of motivational factors on conceptual change in a digital learning context using the dual-situated learning model. *International Journal of Science Education, 32,* 1853–1875. https://doi.org/10.1080/09500690903219156

Veermans, K., van Joolingen, W., & de Jong, T. (2006). Use of heuristics to facilitate scientific discovery learning in a simulation learning environment in a physics domain. *International Journal of Science Education, 28,* 341–361. https://doi.org/10.1080/09500690500277615

Vosniadou, S., & Brewer, W. F. (1994). Mental models of the day/night cycle. *Cognitive Science, 18*(1), 123–183. https://doi.org/10.1016/0364-0213(94)90022-1

Waight, N., & Abd-El-Khalick, F. (2012). Nature of technology: Implications for design, development, and enactment of technological tools in school science classrooms. *International Journal of Science Education, 34*(18), 2875–2905. https://doi.org/10.1080/09500693.2012.698763

Wang, C., Ke, Y.-T., Wu, J.-T., & Hsu, W.-H. (2012). Collaborative action research on technology integration for science learning. *Journal of Science Education and Technology, 21,* 125–132. https://doi.org/10.1007/s10956-011-9289-0

Wu, H.-K., & Huang, Y.-L. (2007). Ninth-grade student engagement in teacher-centered and student-centered technology-enhanced learning environments. *Science Education, 91,* 727–749. https://doi.org/10.1002/sce.20216

Zacharia, Z. C., Olympiou, G., & Papaevripidou, M. (2008). Effects of experimenting with physical and virtual manipulatives on students' conceptual understanding in heat and temperature. *Journal of Research in Science Teaching, 45,* 1021–1035. https://doi.org/10.1002/tea.20260

Zhang, B., Looi, C.-K., Seow, P., Chia, G., Wong, L.-H., Chen, W., et al. (2010). Deconstructing and reconstructing: Transforming primary science learning via a mobilized curriculum. *Computers & Education, 55,* 1504–1523. https://doi.org/10.1016/j.compedu.2010.06.016

Zhang, J., Chen, Q., Sun, Y., & Reid, D. J. (2004). Triple scheme of learning support design for scientific discovery learning based on computer simulation: Experimental research. *Journal of Computer Assisted Learning, 20,* 269–282. https://doi.org/10.1111/j.1365-2729.2004.00062.x

Zheng, R. Z., Yang, W., Garcia, D., & McCadden, E. P. (2008). Effects of multimedia and schema induced analogical reasoning on science learning. *Journal of Computer Assisted Learning, 24*, 474–482. https://doi.org/10.1111/j.1365-2729.2008.00282.x

Zydney, J., & Grincewicz, A. (2011). The use of video cases in a multimedia learning environment for facilitating high school students' inquiry into a problem from varying perspectives. *Journal of Science Education and Technology, 20*, 715–728. https://doi.org/10.1007/s10956-010-9264-1

Chapter 3
Inquiry-Based Science Education: Characterization and Approaches for Use of Information and Communication Technology

3.1 Introduction

Scientific inquiry involves a variety of skills that scientists use to understand the natural world. To help students gain some of these inquiry skills, researchers and teachers prepare different curricula, a variety of instructional resources and diverse teaching strategies. Together, these actions are designed to improve the quality of science teaching.

Designing a teaching methodology that breaks with the linear style of traditional education (transmissive teaching) entails creating strategies that make students think, research, select information, collect evidence, organise arguments and present their conclusions. It is not always easy for students to develop these strategies for a number of reasons: an emphasis on traditional teaching, use of excessive mathematical formalism, inadequate teacher training and a lack of adequate teaching material (Cachapuz, Gil-Pérez, de Carvalho, Praia, & Vilches, 2011), among others.

Furthermore, there are also difficulties regarding changing the perceptions that students have of their role in the classroom; they tend to regard tasks that involve them in a more active role as a type of game, in which teachers act as supervisors and moderators rather than transmitters of knowledge (Bossler, Baptista, Freire, & do Nascimento, 2009).

A current trend in science teaching (with the focus placed more on the students and less on the teacher) is *inquiry-based science education* (IBSE) (National Research Council [NRC], 2000) supported by Information and Communication Technology (ICT) (Ebenezer, Kaya, & Ebenezer, 2011). According to these authors:

> There are two standards pertinent to the use of technologies in scientific inquiry. Using a variety of technologies for investigation refers to the necessary tools (e.g., hand tools; measuring instruments and calculators; electronic devices; and computers for the collection, analysis, and display of data). The use of mathematical tools and statistical software refers

© The Author(s), under exclusive license to Springer Nature Switzerland AG 2019
G. W. Rocha Fernandes et al., *Using ICT in Inquiry-Based Science Education*, SpringerBriefs in Education, https://doi.org/10.1007/978-3-030-17895-6_3

to applying these to collect, analyze, and display data in charts and graphs and to conduct statistical analyses. (Ebenezer et al., 2011, p. 95)

Information and Communication Technologies (ICTs) are used in many sectors, such as engineering, medicine and the financial system. In this chapter, ICT will be referred to as possible "technological tools that support digital resources" for use in teaching. We can also use the term "ICTE", that is, Information and Communication Technologies for Education (ICTE), which include the different digital tools that can be used in education and teaching (ICTE = ICT + Education) (Charlier, Peraya, & Collectif, 2007). By making students an active element in the process of knowledge acquisition, the use of ICTE can help them perform studies that include formulation of research questions, developing hypotheses, collecting data and reviewing the theory (Rutten, van Joolingen, & van der Veen, 2012).

Given the above, the objective of this chapter is to present a discussion on the main ICTE tools used in IBSE and to propose a systematic approach to the main steps that characterise *inquiry activities in science teaching* and their approaches to the use of ICT. As such, our aim is to answer the following questions:

(1) What are the primary ICTs that are used in IBSE?
(2) How can we characterise the main steps of IBSE when they are supported by different ICTs?

To help in our understanding of these questions, we will present a set of studies that discuss the use of IBSE supported by ICTE and that present examples of empirical studies that include results from the use of educational technology in inquiry-based teaching.

This chapter begins with a discussion of the origins of IBSE and inquiry activities in science teaching and tries to find a concept of IBSE that will help us answer the questions proposed in this chapter. Finally, we try to summarise what is currently known regarding the primary ICTs that are used in IBSE and how we can characterise the steps that comprise the inquiry activities supported by ICT.

3.2 The Origins of the Inquiry-Based Approach and Inquiry Activities in Science Education

Some studies report that the first examples of bringing scientific inquiry into the classroom date to the nineteenth century, mainly using school laboratories (NRC, 2000; Zompero & Laburú, 2011). Until the turn of the century, the concept of inquiry was based on the repetition of actions adopted by scientists through the use of a scientific method and a neutral science.

It was only in the early twentieth century that the idea of IBSE gained strength, particularly as a result of the work of Joseph Schwab and John Dewey (NRC, 2000; Trópia, 2011; Zompero & Laburú, 2011). In 1938, John Dewey published *Logic: The Theory of Inquiry*. For Dewey, knowledge begins with a problem and ends with the resolution of the problem. In other words, the act of inquiry requires a systematic approach that involves forming questions, inquiring and concluding and therefore differs from mere definition or demonstration (Dewey, 2007).

In the 1950s, 1960s and early 1970s, the *inquiry-based approach* label was used for most of the curricular projects supported by the National Science Foundation (NSF) in the United States.

In the 1960s, in an attempt to win the space race, the United States of America (USA) invested in human and financial resources in education, helping to produce the so-called first generation of projects for physics (Physical Science Study Committee – PSSC), chemistry (Chemical Bond Approach – CBA), biology (Biological Science Curriculum Study – BSCS) and mathematics (School Mathematics Study Group – SMSG) education in high school. These projects emphasised the use of experimental studies and teaching guided by programmed instruction. The scientific method was divided into clearly demarcated steps: identifying problems, establishing hypotheses for solving them, organisation and execution of experiments to verify the hypotheses and forming conclusions, i.e. validating or refuting the hypotheses. This movement had strong support from professional scientific associations, universities and renowned scholars and was supported by the government, and it subsequently influenced other countries (e.g. England then funded projects through the *Nuffield Foundation*).

According to Trópia (2011), during the 1980s, associations between the *inquiry-based approach* and the theoretical notions of the time drawn from the work of Piaget were formed and applied in science education studies; examples include *alternative conceptions* and *conceptual change* (Posner, Strike, Hewson, & Gertzog, 1982), *science literacy* (NRC, 1996) and the *Science-Technology-Society* (STS) *movement* (Gil Pérez & Castro, 1996). These diverse ideas contributed to the curricular reforms that occurred in the United States and England in the late 1980s and early 1990s. The reforms brought in *inquiry* as part of an approach to science teaching that was contextualised in the reality of the student as opposed to being a mere reproduction of the scientific method (NRC, 1996; Trópia, 2011; Zompero & Laburú, 2011).

The *American Association for the Advancement of Science* (AAAS) founded Project 2061 in 1985 to help all Americans to achieve adequate training in science, mathematics and technology. Starting with its first publication in 1989, "Science for All Americans[1]", Project 2061 has laid down recommendations that state what students should know or be able to do in science, mathematics and technology (AAAS, 1990). The document "Science for All Americans" became the starting point for the national movement for *standards* in science in the 1990s. In 1993, another document, entitled "Benchmarks for Science Literacy: Project 2061", was published; this document provided curricular guidelines for science teaching[2].

In 1996, one of the main documents on the *inquiry-based approach*, the *National Science Education Standards* (NRC, 1996), was published in the United States. This

---

[1] Available from http://www.project2061.org/publications/sfaa/online/sfaatoc.htm

[2] Available from http://www.project2061.org/publications/bsl/online/index.php?home=true

document proposes some guidelines (*standards*) for *science literacy* and acknowledged the importance of IBSE (Zompero & Laburú, 2011). The *standards* for science teaching in the United States are characterised by a detailed level of content that is illustrated with examples, objectives, principles, suggestions and lists of abilities (NRC, 1996).

The document *Inquiry and the National Science Education Standards: A Guide for Teaching and Learning* (NRC, 2000)[3] was published in 2000. It presented the main themes that should be covered as part of *inquiry-based teaching and learning*. These themes are (1) inquiry in science and in classrooms, (2) preparing teachers for inquiry-based teaching, and (3) supporting inquiry-based teaching and learning.

In 2012, the *Framework for K-12 Science Education* was published with the aim of making science education more uniform across the country (NRC, 2012).[4] In 2013, a new set of national standards, *Next Generation Science Standards (NGSS)*, were presented to the American community (NGSS Lead States, 2013).[5]

The projects implemented in the United States, followed by curriculum reform, led the science academies of several countries to rethink their science curricula. According to Boaventura, Faria, Chagas, and Galvão (2011), many of these documents defend the need to develop students' opinions of scientific activity. This can be done using an inquiry-based approach that emphasises problem-solving and critical thinking as early as possible in a real-world context. In this respect, there has been some debate about what students at different levels of education can do with scientific activities, and recommendations have increasingly moved away from the traditional forms of *inquiry* that were advocated in the 1950s and 1960s through the American and English curricular projects.

The official documents of the US reforms, such as those from the AAAS (1990) and the NRC (1996, 2000), are the primary references used in most studies of the *inquiry-based approach* (Bell, Urhahne, Schanze, & Ploetzner, 2010; Olympiou & Zacharia, 2012; Smetana & Bell, 2012; Webb, 2005; Zacharia, Olympiou, & Papaevripidou, 2008). These documents also influenced the implementation of IBSE in the science curricula of various countries. For example, in Australia, the *National Statement on Science for Australian Schools* gave high priority to the development of inquiry and problem-solving skills that can be developed by inquiry work. Such work gives students the opportunity to practice the skills of defining a research problem, formulating hypotheses, designing experiments and analysing and interpreting data (Lin, Hong, Chen, & Chou, 2011). In the United Kingdom, *scientific inquiry* has been defined as a positive proposal for use in science education by the *National Curriculum Orders* and has been strongly supported by the countries members of the *Royal Society* (Lin et al., 2011).

In Singapore, the science curriculum was revised and updated in 2008 to include the concept of *inquiry* as a central and guiding philosophy (Tan & Wong, 2012). Scientific education in Singapore emphasises acquisition of scientific knowledge,

---

[3] Available from http://nap.edu/catalog.php?record_id=9596

[4] Available from https://www.nap.edu/download/13165#

[5] Available from https://www.nap.edu/download/18290#

processes and attitudes as a method for helping students see science as significant and useful (Tan & Wong, 2012). "Science as inquiry is identified as a means for scientific knowledge, issues, and questions to be addressed. The choice of inquiry practiced is dependent upon the context as well as the abilities and readiness of the learners" (Tan & Wong, 2012, p. 198).

Germany responded to the low performance of its students in the 2001 *Programme for International Student Assessment* by introducing the *National Science Education Standards*, which covered four main areas of competence: *domain-specific knowledge*, *methodological knowledge*, *communication* and *judgement* (Bell et al., 2010). *Methodological knowledge* is an area that includes many learning activities that are connected with inquiry-based science and emphasises the importance of this educational dimension in the science curriculum.

Meanwhile, in France, in addition to the project *La Main à la Pâte* which was started in 1996 (Charpak, 1999), in 2011, the Ministry of Education and Development introduced the *Study Programme: Science and Technology* "years 6 to 8". This programme was made up of two parts: a theoretical framework and a syllabus. The theoretical framework contains a set of standards that are designed for use by teaching professionals such that students can achieve general standards of learning; these include the didactic principles that compose the inquiry process and the main skills that are associated with inquiry and technology-based activities. A specific result of learning is stated to be "understanding a problem, planning a scenario, implementing it and analysing and evaluating the solution" (Ministère de l'Éducation et du développement de la petite enfance, 2011). So as not to repeat the scientific method of the 1960s, the programme provides guidance on oral and written communication, reading and the role of attitudes and values held by students.

Portugal also introduced various changes to the science curriculum, starting in 2001 with the Law no. 6/2001 of 18 January, that were driven by a view that the system should be organised around learning competencies (Galvão, Reis, Freire, & Almeida, 2011). The curriculum emphasises a constructivist approach to teaching and learning and places value on the development of critical thinking strategies, the creation of environments for learning inquiry and promoting self-management in learning based on problem-solving and decision-making skills (Galvão et al., 2011).

In Brazil, this trend intensified mainly in the late 1990s with the "National Curriculum Parameters" (Brasil, 1999) and their later additions (Brasil, 2002); with the work of de Carvalho and Collectif (2004, 2013) and de Carvalho, Vannucchi, Barros, Gonçalves, and Rey (2010); and also with the work of the Scientific and Cultural Dissemination Centre (Centro de Divulgação Científica e Cultura – CDCC/USP), which, since 2001, has participated in the programme ABC in Science Education – Hands On (Schiel, Orlandi, & Collectif, 2009).

Although the importance of inquiry-based teaching in science is very clear, its inclusion in official documents in a variety of countries has made its definition rather unclear. In the next section, we will see that over the course of a decade, science teachers have debated, agreed and disagreed on the definition of "inquiry" (Bell et al., 2010; Lucero, Valcke, & Schellens, 2012).

3.3   Establishing the Boundaries of IBSE

With the United States returning to IBSE – mainly during the 1990s as a result of the curricular reform (NRC, 1996, 2000) – we saw in the previous section that other countries also started to adopt the same ideas.

The importance of *inquiry-based learning* is widely recognised. However, while it is not impossible, it is difficult to give a commonly accepted definition (Bell et al., 2010). The term has different meanings in different contexts (Bell et al., 2010), and this is also the case in the science curricula; these differences reflect the political, economic and societal needs of each country (Trópia, 2011; Zompero & Laburú, 2011). These meanings range from the traditional concepts contained in the curriculum reforms of the 1950s and 1960s, which promoted the use of the scientific method in the classroom, to the notions that seek to go beyond this view, which introduce discussions about the nature of science (NOS) and other dimensions that have been part of the conditions for performing science since the 1980s (Galvão et al., 2011; NRC, 1996; Trópia, 2011).

What do we currently mean when we talk about IBSE and *Inquiry-based Activities in Science Education (IBASE)*? What we can say is that inquiry activities are not currently performed using closed steps, i.e. by directing students to perform them in an algorithmic manner, as in a supposed scientific method (Zompero & Laburú, 2011) – a method that has been subject to the criticism of a number of science education researchers (Cachapuz et al., 2011).

Inquiry-based teaching is no longer focused on training scientists, as it was in the 1960s. Currently, *inquiry-based teaching* is used for other purposes, such as promoting *scientific literacy* (Sasseron & Carvalho, 2011); developing cognitive skills in students (Lee, Linn, Varma, & Liu, 2010); forming hypotheses, recording results, analysing data and developing the ability to construct arguments (Chen & Looi, 2011; de Carvalho & Collectif, 2004, 2013; Driver, Newton, & Osborne, 2000); reflecting on socio-scientific questions (Barab, Sadler, Heiselt, Hickey, & Zuiker, 2007; Cher Ping, 2008; Nelson, 2007); discussing the NOS and the role of the scientist (Driver, Leach, Millar, & Scott, 1996); and developing experience with specific software and programs (Clark & Sampson, 2007; Webb, 2005), among other purposes. In the search for a concept and given the need to draw some defining lines around it, there is a tangible sense of confusion in the literature regarding the terms *inquiry-based learning*, *inquiry-based teaching* and *inquiry-based activities*.

We have brought together some works that use the concept of IBSE as a theoretical framework. Table 3.1 presents a summary of the main elements that characterise the concept.

Table 3.1 indicates that *inquiry* appears less often cited as a method (Tan & Wong, 2012), content (NRC, 2000) or teaching strategy (Bossler et al., 2009). Instead, it is more often characterised by inquiry activities, development of steps and procedures, learning approaches and development of skills.

Many studies adopt the definition given by the NRC (1996, 2000, 2012), Barab et al. (2007), Ketelhut (2007), Kubasko et al. (2008), Tan and Wong (2012),

Table 3.1 Identifying the main features of the concept of "inquiry"

Definition of inquiry-based science education	
Keywords	Authors
Method	NRC (1996), Tan and Wong (2012)
Activities	de Carvalho and Collectif (2004, 2013), Lucero et al. (2012), NRC (1996, 2000), Oh (2010), Tan and Wong (2012), Ucar and Trundle (2011)
Steps and procedures	Bossler et al. (2009), Jaakkola and Nurmi (2008), Kawalkar and Vijapurkar (2013), Lucero et al. (2012), NRC (1996), Waight and Abd-El-Khalick (2007), Zompero and Laburú (2011)
Ways to learn science Learning approaches	Barab, Sadler, Heiselt, Hickey, and Zuiker (2007), Dori and Sasson (2008), Jaakkola and Nurmi (2008), NRC (2000), Scalise et al. (2011)
Development of skills (cognitive and manipulative)	Dori and Sasson (2008), NRC (2000), Tan and Wong (2012), Zompero and Laburú (2011)
Content	NRC (2000)
Teaching strategy	Bossler et al. (2009)
Concepts of NRC (1996) and NRC (2000)	Barab et al. (2007), Ketelhut (2007), Kubasko, Jones, Tretter, and Andre (2008), Tan and Wong (2012), Tolentino et al. (2009), Waight and Abd-El-Khalick (2007)

Tolentino et al. (2009) and Waight and Abd-El-Khalick (2007). The NRC itself (1996) describes various features of the *inquiry-based approach*; i.e. it defines inquiry as *activities*:

> Inquiry also refers to the activities of students in which they develop knowledge and understanding of scientific ideas, as well as an understanding of how scientists study the natural world. (NRC, 1996, p. 23)

In addition to defining *inquiry* as a "multifaceted activity", it also refers to it as a series of *steps and procedures*:

> Inquiry is a multifaceted activity that involves making observations; posing questions; examining books and other sources of information to see what is already known; planning investigations; reviewing what is already known in light of experimental evidence; using tools to gather, analyze, and interpret data; proposing answers, explanations, and predictions; and communicating the results. (NRC, 1996, p. 23)

As a starting point, we will not assume that IBSE and IBASE are necessarily the same thing, although they are related. IBSE is a method of teaching and learning science (Barab et al., 2007; Dori & Sasson, 2008; Jaakkola & Nurmi, 2008; NRC, 2000; Scalise et al., 2011). Inquiry activities are part of this teaching-learning environment, are designed to help develop certain cognitive skills and are characterised by specific steps and procedures. These procedures are made up of three phases: (a) planning (before), (b) development (during) and (c) reflection (after).

(a) *Planning inquiry activities*: in planning an inquiry-based activity as part of "IBSE", the teacher should be clear about the content of the work to be performed (NRC, 2000), the cognitive and manipulative skills developed by the students (Dori & Sasson, 2008; NRC, 2000; Tan & Wong, 2012; Zompero & Laburú, 2011) and the steps and procedures that make up the activity (Jaakkola & Nurmi, 2008; Kawalkar & Vijapurkar, 2013; Lucero et al., 2012; NRC, 1996; Waight & Abd-El-Khalick, 2007).

The inquiry activities that form the *inquiry* phase can be classified into two types: *object-on* and *mind-on*. Activities that are *object-on* can be classified into two types: (1) *hands-on*, in which students manipulate a real object and perform an experiment to obtain an answer to a given inquiry problem, and (2) *ICT-on*, in which students interact with ICTE to obtain results from an inquiry study. Examples include using data from the Internet (Ucar & Trundle, 2011; Van Zee & Roberts, 2006; Waight & Abd-El-Khalick, 2007), using simulations (Bell & Trundle, 2008; Lin, Hsu, & Yeh, 2012) and virtual laboratories (Dori & Sasson, 2008; Jaakkola & Nurmi, 2008) and even using a *Game* in cases that relate to socio-scientific questions (Nelson, 2007; Tolentino et al., 2009). Activities of a *mind-on* type are designed to develop the critical thinking of the students, not only the physical component of learning (hands-on) (Donnelly, O'Reilly, & McGarr, 2012). Addressing socio-economic questions is one example of mind-on activities (Furberg & Ludvigsen, 2008; Zydney & Grincewicz, 2011).

(b) *Developing inquiry activities*: this is the phase that is characterised by the use of steps and procedures in IBSE. A number of authors provide a series of steps for performing scientific activities (de Carvalho & Collectif, 2004, 2013; Ebenezer et al., 2011; Jaakkola & Nurmi, 2008; NRC, 2000; Rutten et al., 2012). The students start with a question or a problem (provided by the teacher or invented by themselves), formulate hypotheses, perform actions (studies) to answer the problem, collect and organise the data and respond to the questions. Developing activities using steps is much more than applying a scientific method because it involves the development of more complex skills (hands-on as well as mind-on): "When engaging in inquiry, students describe objects and events, ask questions, construct explanations, test those explanations... and communicate their ideas" (NRC, 1996, p. 2).

(c) *Reflecting on the inquiry activities*: when an inquiry-based activity comes to an end, the teacher encourages the students to reflect on the actions that they performed. This is a mind-on skill in which the students discuss the process and share their doubts, difficulties and solutions. We can say that this is the point at which "Inquiry requires identification of assumptions, use of critical and logical thinking, and consideration of alternative explanations" (NRC, 1996, p. 23).

It is important to note that the development of inquiry activities within the perspective of the *inquiry-based approach* should not be centred in a closed-minded attitude towards science. The inquiry process should go beyond the instrumentalist technical activities. For example, the process may include discussing the social and political implications and relations of scientific inquiry in society, including the controversies and limits of science that arise when performing the activities (Cachapuz et al., 2011; Trópia, 2011).

3.4   The Main ICT Tools Used in Inquiry Activities

Studies in which students develop inquiry activities supported by educational technologies have gained popularity in recent years. In this chapter, we present some studies that used simulations and simulation software, virtual laboratories, hypermedia, multimedia, the Internet, remote laboratories, smartphones, tablets and other digital technologies (*objects-on activities*) as support for inquiry activities in teaching/learning situations. Table 3.2 characterises the main educational technologies, the primary studies that use "IBSE" as the theoretical framework and the effects of using the technologies.

The use of digital technologies together with IBSE elicits a more collaborative style of teaching that is centred on the student and removed from the idea of the "traditional scientific method". The effects that have been found in the studies and summarised in Table 3.2 are central to the development of this type of teaching: scientific reasoning, conceptual evolution, motivation, engagement, thinking skills, encouraging scientific argumentation, changes in attitudes in regard to teaching and science, learning, collaborative work and others. In what follows, we present explanations of the ICT-on inquiry activities presented in Table 3.2, with the aim of characterising the main types of work that use some form of educational technology in inquiry-based science teaching.

3.4.1   Inquiry Activities that Use Hypermedia, Multimedia and the Internet

As can be observed from Table 3.2, the use of the *Internet with its media and hypermedia support* has several important characteristics from the perspective of IBSE. Internet-based science learning has been studied for more than a decade. We can cite the work of Lee et al. (2011), who performed an important review of articles published in periodicals on the subject of *Internet-based science learning* for the period 1995–2008. This review details some conclusions found from performing inquiry activities using the Internet. For example, maintaining control over the student is essential for improving their attitudes and motivation for learning science through the Internet. At the same time, appropriate supervision of teachers, moderators or learning environments using the Internet is also essential for Internet-based science learning.

The study by Gelbart et al. (2009) presents the results of the impact of web-based *inquiry* on students' understanding of *genetics*. Clark and Sampson (2007) investigated personally seeded discussions (PSD) to scaffold online argumentation. Hoffman et al. (2003) studied the nature of the understanding of scientific content, inquiry use and strategies of 830 sixth-grade students using the Internet (via Artemis, an interface to the Digital Library) to research *astronomy*, *ecology*, *geology* or *weather*. The data analysis demonstrated that students accumulated significant

Table 3.2 The primary educational technologies available for use in performing inquiry activities and their characteristics and effects

ICT-on inquiry activities	Studies	Summary of features of "inquiry"	Effects
Hypermedia, multimedia, Web and Internet	Dori, Tal, and Peled (2002); Gelbart, Brill, and Yarden (2009); Hoffman, Wu, Krajcik, and Soloway (2003); Kubasko et al. (2008); Lee et al. (2011); Mistler-Jackson and Butler Songer (2000); Shin, Jonassen, and McGee (2003); Songer, Lee, and Kam (2002); So (2012); Ucar and Trundle (2011); van Zee and Roberts (2006); Varma and Linn (2012); Waight and Abd-El-Khalick (2007, 2011); Zydney and Grincewicz (2011); Clark and Sampson (2007)	*Web*: search for web pages; problem-solving; sharing of data available on the Web; guided and open inquiry; search strategies; synchronous and asynchronous communication; collaborative work; collaboration with scientists *Multimedia-hypermedia*: dynamic and static representation of scientific phenomena; ubiquitous visualisation of phenomena; problem solving; guided content	Need for actions and support by the teacher; reduction in teaching time; attitude, scientific reasoning, conceptual change, motivation and engagement of the student; self-effectiveness; development of critical thinking
Simulations and software simulations	Bell and Trundle (2008); Lin, Hsu, and Yeh (2012); Quellmalz, Timms, Silberglitt, and Buckley (2012); Rutten et al. (2012); Scalise et al. (2011); Smetana and Bell (2012); Snir, Smith, and Raz (2003); Stieff (2011); Zacharia (2003); Clark and Sampson (2007)	Representation of scientific phenomena; problem-solving; experimentation, navigation and manipulation of variables; observation of changing phenomena; construction and analysis of models representing different phenomena	Encouraging scientific arguments; developing ideas and the conceptual evolution; active engagement; conceptual and analytical understanding; complementing traditional teaching; identifying and correcting misconceptions; construction of different representations

(continued)

Table 3.2 (continued)

ICT-on inquiry activities	Studies	Summary of features of "inquiry"	Effects
Virtual laboratories	Donnelly, McGarr, and O'Reilly (2011); Donnelly et al. (2012); Dori and Sasson (2008); Jaakkola and Nurmi (2008); Jaakkola, Nurmi, and Veermans (2011); Olympiou and Zacharia (2012); Pyatt and Sims (2012); Zacharia et al. (2008)	Combining virtual resources with real experiments; representations of scientific phenomena; solving problems; manipulating variables; modelling	A preference for using the virtual laboratory; conceptual understanding; teaching support; conceptual understanding and graphical skills; interest and motivation; individualised learning
Laboratories and remote data	Kong, Yeung, and Wu (2009)	Researching data; visualising scientific phenomena; problem-solving; interpreting variables; manipulation and control of real equipment; sharing resources between different institutions; data from distance-based research; collaborating with scientists	Enthusiasm shown by students; ubiquitous access
Computer games and serious games	Squire and Jan (2007)	Prediction-observation-explanation strategy; problem-solving; manipulation of variables	Logical reasoning; gains in knowledge and impact on learning; changing attitudes; scientific thinking and argumentation
Multi-user virtual environments (MUVE)	Barab et al. (2007); Cher Ping (2008); Furberg and Ludvigsen (2008); Hakkarainen (2003); Ketelhut (2007); Lin, Wang, and Lin (2012); Nelson (2007); Tolentino et al. (2009)	Questioning, reflecting and developing discourse on conceptual, socio-scientific and ethical issues in science; the process of scientific inquiry and collaborative learning	Active engagement; motivation; complex system thinking skills; producing socio-scientific discourse

(continued)

Table 3.2 (continued)

ICT-on inquiry activities	Studies	Summary of features of "inquiry"	Effects
Computer-assisted instruction (CAI)	Barak and Dori (2011), Ebenezer et al. (2011), Hansson, Redfors, and Rosberg (2011)	Guided tutorial system; problem-solving; describing scientific phenomena; modelling activities; analysis, synthesis and interpretation using tests	Improvements in performance, research and planning capabilities; attitudes related to science; a reduction in alternative conceptions; an improvement in the understanding of concepts and theories, scientific arguments, modelling and collaborative work
Photos, videos, wikis, chats	Kim and Herbert (2012); Kim, Miller, Herbert, Pedersen, and Loving (2012); Wang, Ke, Wu, and Hsu (2012)	Chat as a means of exchanging information; videos as a way of representing and explaining scientific phenomena; photographs as a static record and resource for mapping and analysing phenomena	Photographs as a tool that explores ideas and encourages scientific arguments; videos as an aid to questioning; creating meanings (significance) and promoting changes in ideas
Smartphones and tablets	Looi et al. (2011), Zhang et al. (2010)	Instruments for mobile learning and sharing of data; tasks involving querying and collecting data and working in groups	More personalised learning; positive attitudes in regard to science teaching

understanding using online inquiry, although the accuracy and depth of their understanding varied. The results suggest that students can develop more accurate and deeper understanding if they use searches and evaluate their strategies appropriately, if the resources are chosen wisely and if support for the learning environment is widely available. These actions are most often associated with the role of the teacher in accompanying the development of the students' work.

This chapter also revealed a series of science projects in which the inquiry process is based on sharing data across a network (Hoffman et al., 2003; Kubasko et al., 2008; Mistler-Jackson & Butler Songer, 2000). In these projects, a group of individuals – students, teachers or scientists – share data and collaborate to investigate scientific questions and current events. It is important to state that the manner in which data sharing activities are performed can vary widely, including in terms of the target audience, objective, use of technological tools, autonomy of the teacher and flexibility of the data collection standards.

3.4.2   Inquiry Activities that Use Simulations and Simulation Software

Inquiry-based teaching and learning can also be supported through the use of simulations and educational software (Lindgren & Schwartz, 2009; Scalise et al., 2011; Smetana & Bell, 2012). This area has become a subject of study for many researchers in science teaching and psychology (Mayer, 2009; She & Chen, 2009). We chose to analyse simulation packages and educational software separately from virtual laboratories because each of these tools can have distinct usages.

Notwithstanding the difficulties involved in bringing ICT into the educational context, some studies have been performed with the aim of understanding and recording the main trends in inquiry-based science teaching supported by use of simulations. For example, in the review by Rutten et al. (2012), the authors compare the teaching conditions with and without simulations. The findings demonstrate positive results when simulations were used as a substitute for or to improve traditional teaching, especially when inquiry activities are employed prior to entering the physical laboratory (pre-laboratory activities).

Scalise et al. (2011) reviewed 79 studies that are related to the use of computer simulations for science students between grades 6 and 12 in American schools. Thirty-nine studies demonstrated learning gains that were associated with *simulations supporting scientific inquiry*. According to these authors, the simulations help students develop research questions, design experiments, set up simulation projects and obtain and analyse simulation data (Scalise et al., 2011).

Another work that is part of our study is the review performed by Smetana and Bell (2012) of the impact of computer simulations on the teaching and learning of science; this work also summarises what is currently known about IBSE and provides suggestions for future studies. The main results of this review suggest that science teaching using simulations can often be more effective than traditional pedagogical practices (based on lectures, the use of textbooks and physical experiments) and play a supporting role in building scientific knowledge, developing skills (e.g. researching, collecting data and manipulating variables) and promoting increased understanding of scientific concepts.

> Computer simulations provide many advantages to support calls for inquiry-based, learner and knowledge-centered instruction (NRC, 1996). For example, simulations offer the advantage of flexibility, promoting students' active engagement in problem-solving and higher-order thinking and reinforced practice. (Smetana & Bell, 2012, p. 1338)

The results of the studies presented in Table 3.2 demonstrate that a well-designed computer simulation, used as part of an inquiry-based teaching model, can significantly help students develop scientific understanding, thinking skills, scientific argumentation, development of ideas, conceptual evolution and active engagement, for example.

3.4.3 Inquiry Activities that Use Virtual Laboratories

A summary of the characteristics of the use of virtual laboratories as inquiry activities and the effects of virtual laboratories is given in Table 3.2. These characteristics include combining virtual resources with real experiments, representing scientific phenomena, solving problems and manipulating variables.

Over the last few decades, various studies have tried to understand and document the value of *hands-on* inquiry activities using *physical experiments* (PEs) and, more recently, *ICT-on* education supported by *virtual experiments* (VEs) (Jaakkola et al., 2011; Jaakkola & Nurmi, 2008; Pyatt & Sims, 2012; Russell, Lucas, & McRobbie, 2004; Zacharia, 2007; Zacharia et al., 2008). For us, comparative studies have been performed to determine which of these two types of experiments (PEs or VEs) have been used to develop inquiry activities. Many science teachers question whether it is better to combine PEs with VEs or to use the resources separately. Jaakkola and Nurmi (2008) investigated what would be the best choice for teaching *simple electricity* concepts to students in the fifth grade. The results indicated that a combination of VEs and PEs resulted in statistically significant learning gains compared with using only VEs or PEs and also stimulated the conceptual understanding of the students in a more efficient manner.

Zacharia et al. (2008) confirmed the results of Jaakkola and Nurmi (2008). These researchers found the use of VM to be the differentiating factor, although not all studies indicate the same results when using a combination of real and virtual experiments. For example, in the study by Pyatt and Sims (2012), university chemistry students that performed an inquiry-based *stoichiometry* activity in the virtual laboratory obtained the same results as students that used physical equipment and materials. There was no statistically significant difference between the mean assessment results of the virtual and physical laboratory groups.

In a later study, Olympiou and Zacharia (2012) performed an experiment using a pre- and post-study comparison methodology under three conditions: 23 students used PEs, 23 students used VEs and 24 students used a combination of VEs and PEs. The results demonstrated that using a combination of PEs and VEs improved the students' conceptual understanding in the area of *light and colour* compared with the use of only PEs or VEs. These results are similar to those of Zacharia et al. (2008), although the study does have limitations. These limitations are related to the small number of participants, use of a specific group of subjects (undergraduate students), analysis of only one specific topic (light and colour) and use of only one data source (conceptual tests). We believe that this study could have had better student learning results if it had used more data sources and focussed on the process, not only the final results.

Considering the studies listed in Table 3.2, virtual experimentation offers many potential gains in learning that can contribute to *IBSE*: it removes spatial and temporal restrictions on the students, it is low-cost, it is easy to access, and, as a rule, it moves the centre of learning from the teachers to the students (Donnelly et al., 2011). Despite these advantages, including ICT in an inquiry-based approach is a complex process of change that requires careful analysis by researchers, educators and teachers, given that not all of them have the knowledge to do so (Donnelly et al., 2011).

3.4.4 Inquiry Activities that Use Laboratories and Remote Data

Inquiry activities that use remote-controlled experiments performed via the Internet allow students to manipulate or control real equipment and conduct scientific research at a distance using specific hardware and software. The work of Kong et al. (2009) assessed the level of student learning after the students used LabVNC (an open-source remote-controlled software package) and recorded the opinions of the teacher regarding the use of LabVNC in science teaching. The results of this study demonstrate that the positive opinion of the teacher regarding the pedagogical value of remote-controlled experiments and the enthusiasm of the students when using LabVNC reflect the potential for using it to perform inquiry activities. These results are consistent with those of Lowe, Newcombe, and Stumpers (2012), who described tests that used remote laboratories within secondary schools and studied the reactions of 112 students and teachers as they interacted with the laboratories.

Studies have demonstrated that when a remote laboratory is used appropriately, it can provide several potential benefits. These include the abilities to share resources among various institutions, improve accessibility to equipment that would otherwise not be available due to cost or technical reasons and help increase experimental activity (Kong et al., 2009; Lowe et al., 2012). Whereas the use of remote laboratories is more common in higher education, its role in teaching science at secondary level is still very limited and would benefit from other studies being performed in this area, particularly with respect to its effectiveness.

3.4.5 Inquiry Activities that Use Computer Games and Serious Games

In the literature, there are studies that relate IBSE with *computer games* (Enyedy, Danish, Delacruz, & Kumar, 2012; Squire & Jan, 2007). For example, Squire and Jan (2007) considered whether augmented reality games on portable devices can be used to engage students in inquiry and scientific thinking (particularly argumentation), how the game structures affect student thinking and the impact of role playing on learning. The results demonstrate that these games have the potential to engage students in inquiry activities and in meaningful scientific argumentation about scientific phenomena. However, one must exercise caution when considering the results of the study of Squire and Jan (2007). First, the inquiry activity involved in the game lasted only a short time, much less time than a real inquiry would, which prevented the students from developing more in-depth scientific inquiry. A second limitation is the nature of the *inquiry-based approach*, which lacked the systematic data that would be provided by a pre-test/post-test approach to measuring student performance. A final limitation is the active role that the researchers and facilitators played in supervising the game. The limitations of Squire and Jan (2007) are equally valid for a range of studies that use *computer games* in science teaching.

3.4.6   Inquiry Activities that Use Multi-user Virtual Environments (MUVEs)

More immersive three-dimensional virtual environments, which are known as MUVEs or virtual worlds, have become popular in recent years and have become a focus of attention in education. Virtual worlds, such as *Second Life*, do not have rules and goals and, as such, are distinct from *computer games*. It is important to note that several studies of MUVEs have been conducted to assess their applicability to science teaching (Barab et al., 2007; Dede & Barab, 2009; Ketelhut, 2007). For example, Barab et al. (2007) used a sample of 28 fourth-grade students to investigate the applicability of the *MUVE educational game Quest Atlantis* for producing a socio-scientific narrative and the interactive role of the game for supporting learning in an inquiry process. The MUVE tells the story of a city that is faced with ecological, social and cultural decadence (similar to current global challenges) due to a "blind" search for prosperity and modernisation by the governing bodies (Barab et al., 2007). According to these researchers, the students engaged with the inquiry challenge proposed by the MUVE entered into a rich scientific discourse, presented quality work and learned the proposed science content. In addition, by using the incorporated narrative, the students developed a rich conceptual and ethical understanding of science.

The results of the studies analysed on MUVEs also have limitations. The main limitation centres on the involvement of the teachers while the inquiry activities using the MUVE took place, given that most of the studies reviewed here were performed by the researchers themselves. For example, the study of Cher Ping (2008) gave no information regarding the role of the teachers and their training in using the MUVE. According to these authors, the participant teachers felt that they were not competent and not sufficiently confident working with the MUVE (Cher Ping, 2008).

3.4.7   Inquiry Activities that Use Computer-Assisted Instruction (CAI)

Several studies have considered the use of computers to support science teaching or *computer-assisted instruction* (CAI). This topic refers to including ICT, primarily computers, in science teaching. A range of work has sought to investigate the potential of CAI in science teaching, including a meta-analysis of the effectiveness of CAI in science education by Bayraktar (2001); a study of the conceptual change enabled by CAI from Webb (2005); the tutorial system of Soong and Mercer (2011); a study of "prescriptive tutoring", perceptions of fluency with CAI and levels of scientific inquiry abilities in Ebenezer et al. (2011); and others. The following effects of inquiry activities using CAI are reported in the works of Barak and Dori (2011), Ebenezer et al. (2011) and Hansson et al. (2011): improvements in the performance, capabilities and planning of inquiries; improved attitudes towards science; a reduction in alternative conceptions; improvements in conceptual and theoretical understanding; better scientific argumentation; improved modelling; and increased collaborative work (see Table 3.2).

3.4.8   Inquiry Activities that Use Specific Resources: Photographs, Videos, Wikis and Chats

Also exists in the literature a number of works in which IBSE used specific resources (photographs, videos, wikis and chats), as indicated by Table 3.2. The study by Kim and Herbert (2012) sought to bring together teachers and scientists to establish *communities of inquiry*. The authors present the *Inquiry Resources Collection (IRC)*, which is based on a *wiki* and *chat*. These resources were developed by scientists to help novice science teachers design inquiry-based lessons. For Kim and Herbert (2012), "the collaborative managing and sharing of knowledge in a professional development program via a wiki environment is the key to developing a practical resource for novice teachers teaching scientific inquiry" (p. 504).

In addition, the study by Kim et al. (2012) presents the *Professional Learning Community Model for Entry into Teaching Science (PLC-METS)*. This is an implementation of a program based on *professional learning communities* that uses inquiry activities, through a combination of current scientific research and the use of information technology, to aid scientific understanding and teaching practice in novice science teachers. This *learning community* learns from teacher trainers, researchers, scientists, basic education teachers, educators and others via interaction and cooperation, with the objective of sharing practices, ideas and knowledge (Kim et al., 2012).

3.4.9 Inquiry Activities that Use Mobile Technologies

Studies that focus on the use of smartphones and tablets are related to *mobile technologies* or *learning mobile*. As an example, Looi et al. (2011) sought to test and refine a research project that uses mobile technologies in the third-grade science curriculum of a primary school to study the subject of the *human body* using inquiry activities. The results demonstrate that the class that performed the experiment had better performance than the other classes, which used traditional methods of teaching and assessment. The researchers found that the students in the classes supported by the use of mobile technologies learned science and performed inquiry activities in a more personal, deep and engaging way, and they also developed positive attitudes towards science. The study of Zhang et al. (2010) considered the effects of mobile technologies when 39 primary school students used them daily in science education. The results demonstrated that the students became actively involved in inquiry-based tasks such as collecting data and working in a group.

Having presented a range of digital technologies, it is important to state that the design and use of these tools for developing inquiry activities requires a certain degree of caution. In particular, an excess of resources can impede the learning process. As an example, the study of Waight and Abd-El-Khalick (2007) assessed the impact of a multimedia environment on the representation of scientific inquiry for 42 sixth-grade students. The results indicated that the technology used – microcomputer-based laboratories; simulations and microworlds; telecommunication technologies, including e-mail and Internet interfacing; and accessing and using Web-based databases – acted to restrict rather than promote inquiry in the classroom. When computers were used, the group activities became more structured, with a focus on sharing tasks, and less time was dedicated to group discussion, with a notable drop in meaning-making discourse by students.

In this chapter, we have concluded that there are several possibilities to integrate ICT into IBSE tasks. However, it is up to the teacher to decide on the criteria for selecting the resources and planning the implementation of ICT-based inquiry activities.

3.5 The Main Steps in Inquiry Activities in Science Teaching and their Approaches to the Use of ICT

In the first part of this chapter, we saw that the different definitions of *inquiry* are all conceived as involving numerous procedural and conceptual activities that are based on developing strategies and steps. Such strategies and steps include the following: asking questions; formulating hypotheses; designing experiments; making predictions using equipment, observations and measurements; being concerned with accuracy, precision and errors in the data; recording and interpreting data; assessing test results; checking contradictions or anomalous data; presenting and evaluating arguments; building explanations (for oneself and for others); constructing various representations of the data (e.g. graphs, maps and three-dimensional models); linking theory and practice; performing statistical calculations; making inferences; and revising theories or models (Bell et al., 2010; de Carvalho & Collectif, 2004, 2013; Gil Pérez & Castro, 1996; NRC, 1996, 2000). To review the main studies that analyse the strategies used for performing *inquiry activities* and the use of *educational technologies* and their results, it is first necessary to specify which strategies or steps we are referring to.

The NRC (, 2000) divides the steps of IBSE into five major categories: (1) questions, (2) evidence, (3) explanations, (4) connections and (5) communication. These steps seek to primarily stimulate (1) building scientific knowledge and (2) skills and attitudes of students, thereby encouraging them to search for a deeper understanding of the relationships between what they observe and natural occurrences. For Ebenezer et al. (2011), the three main characteristics of *scientific inquiry* are (1) scientific conceptualisation, (2) scientific investigation and (3) scientific communication. For our purpose, which is to determine the main steps that characterise the development of inquiry activities in science teaching and their possible relationships to the use of *educational technologies*, these categories seem too general. As one possible proposal, we present in Table 3.3 the main inquiry steps that some manner use some sort of ICT.

For us, the IBSE steps occur jointly among students, teachers (who guide the inquiry), classes (which are composed of more than one student), objects (which can be experimental or virtual) and experts (who complement the inquiry). These steps normally follow the process shown in Fig. 3.1, in which the activities are focused on the student; the activities are mediated by the teacher or object; the activities require the object but not necessarily the teacher; and the interaction of the student can occur with the object, class, teacher and/or the expert. The IBSE steps occur jointly among students, teachers (who guide the inquiry), classes (which are composed of more than one student), objects (which can be experimental or virtual) and experts (who complement the inquiry). These steps normally follow the process shown in Fig. 3.1, in which the activities are focused on the student; the activities are mediated by the teacher or object; the activities require the object but not necessarily the teacher; and the interaction of the student can occur with the object, class, teacher and/or the expert. For example, when performing an investigation, the student can use an object (physical or virtual) and ask the teacher for help (S → O → T). The student can use an object and perform the investigation with the class (S → O → C) or even not use an object and interact directly with the class (A → C) or the teacher (A → P) (Fig. 3.1).

The stages presented in Table 3.3 and the elements in Fig. 3.1 are not listed in any fixed order: the students can pass through the steps in the order needed and return to them, if necessary, using the object, teacher, classmates, expert or even individually with the experimental or virtual objects (among those that are part of our study). Analyses of the manner in which the stages are performed indicate that IBSE can adopt a variety of forms (Bell et al., 2010).

From Table 3.3, we can observe various common themes. Many of the authors cited admit that developing a proposal for an inquiry activity requires determining a problem to be analysed, stating hypotheses, performing the inquiry process, interpreting new information and communicating that information (Bell et al., 2010).

The examples in Table 3.3 cover a wide range of stages used in *inquiry activities* and to understand how IBSE works through educational technology. The examples cited in Table 3.3 do not provide a complete overview of all the stages that characterise IBSE, the *inquiry activities* and the use of the educational technologies that are part of this teaching environment. The goal of Table 3.3 is defining a model for the possible stages to be used in an ICT context. In this sense, this is similar to the work of Bell et al. (2010) on collaborative inquiry-based learning and the inquiry framework of So (2012) regarding "*a resource-based e-learning environment for science learning in primary classrooms*".

By compiling a variety of approaches to IBSE, we present a set of five categories that are able to characterise the main ideas behind the steps of IBSE that are

Table 3.3 The main inquiry stages and a possible approach for using ICT

Elements of enquiry activities	Stages	Synthesis of stages	ICT possible
Problem	Making observations (Ketelhut, 2007; NRC, 1996) Exploring the world (Barab et al., 2007) Identifying questions and concepts that guide scientific investigations (Kim et al., 2012) Engagement of learners in scientific questions (Zydney & Grincewicz, 2011) Background research (Shin et al., 2003) Predict (Zacharia et al., 2008)	Explore the world	Videos Web Hypermedia Multimedia MUVE Photos
	Formulating researchable questions or testable hypotheses (Ebenezer et al., 2011; Rutten et al., 2012) Orientation/question (Bell et al., 2010) Generating a research question (Dori & Sasson, 2008; Squire & Jan, 2007) Identifying the study problem (Dori et al., 2002; Scalise et al., 2011)	Present a problem	Simulation Software of simulation
	Making observations to answer the questions (including guarding against perceptual bias) (Squire & Jan, 2007) Reviewing evidence to develop and address the questions (Zydney & Grincewicz, 2011) Defending and challenging claims (Zacharia et al., 2008)	Reflect on the problem	

(continued)

Table 3.3 (continued)

Elements of enquiry activities	Stages	Synthesis of stages	ICT possible
Hypothesis	Demonstrating logical connections between scientific concepts guiding a hypothesis and the design of an experiment (Ebenezer et al., 2011) Formulating hypotheses (Barab et al., 2007; Bell et al., 2010; Dori & Sasson, 2008; Jaakkola & Nurmi, 2008; Ketelhut, 2007; Quellmalz et al., 2012; Rutten et al., 2012; Ucar & Trundle, 2011) Formulating explanations to address the questions (Zydney & Grincewicz, 2011) Writing about the hypotheses (Scalise et al., 2011)	Generate hypotheses	Web Wiki (or Google Docs)
	Gather evidence (Quellmalz et al., 2012) Evaluate hypotheses in light of evidence (Quellmalz et al., 2012)	Evaluate the hypotheses	Simulation Software of simulation Virtual and remote laboratory
Investigative process	Designing and conducting scientific investigations (Bell et al., 2010; Dori et al., 2002; Ebenezer et al., 2011; Kim et al., 2012; NRC, 1996; Quellmalz et al., 2012) Developing methods, tools and rationale to explain results (Squire & Jan, 2007) Examining books and other sources of information to see what is already known; reviewing what is already known in light of experimental evidence (NRC, 1996) Developing theories based on and developing evidence (Squire & Jan, 2007) Studying others' research (Squire & Jan, 2007) Designing experiments (Dori & Sasson, 2008; Scalise et al., 2011) Setting up projects (Scalise et al., 2011)	Plan the inquiry	Web Wiki Mental maps tools

(continued)

Table 3.3 (continued)

Elements of enquiry activities	Stages	Synthesis of stages	ICT possible
	Using measurement instruments to collect date (Ebenezer et al., 2011) Collecting data (Dori et al., 2002; Rutten et al., 2012; Scalise et al., 2011; Shin et al., 2003) Collecting and evaluating evidence (Barab et al., 2007; Ucar & Trundle, 2011) Making discoveries (Barab et al., 2007; Jaakkola & Nurmi, 2008) Conducting investigations (Bell et al., 2010; Dori & Sasson, 2008) Conducting iterative trials (Quellmalz et al., 2012) Getting data and results (Scalise et al., 2011) Observe (Zacharia et al., 2008)	Investigate	Web Simulation MUV Software Virtual and remote laboratory CAI
Analysis and interpretation	Proposing answers, explanations and predictions (NRC, 1996) Evaluating explanations (Zydney & Grincewicz, 2011) Using mathematical tools and statistical software to analyse and display data in charts and graphs (Ebenezer et al., 2011) Rigorously testing and evaluating the plausibility of discoveries in the search for new understanding (Barab et al., 2007; Jaakkola & Nurmi, 2008) Using mathematics to improve investigations and communications (Kim et al., 2012) Gathering and analysing data (Dori & Sasson, 2008; Ketelhut, 2007; Scalise et al., 2011; Shin et al., 2003) Transforming observati[1]ons into findings (Zacharia et al., 2008)	Analyse the data obtained	Calculators Software Virtual and remote laboratory CAI

(continued)

Table 3.3 (continued)

Elements of enquiry activities	Stages	Synthesis of stages	ICT possible
	Recognising how investigation itself requires clarification of research questions, methods, comparisons and explanations and weighing evidence using scientific criteria to find explanations and models (Ebenezer et al., 2011)	Interpret new information	Tools excel Mental maps tools
	Formulating explanations (Barab et al., 2007)		
	Challenging one's understandings (Barab et al., 2007)		
	Formulating, revising and analysing scientific explanations and models using logic and evidence (Kim et al., 2012)		
	Justifying explanations (Zydney & Grincewicz, 2011)		
	Predicting, observing and explaining findings (Quellmalz et al., 2012)		
	Noisy data and error analysis (Scalise et al., 2011)		
	Synthesising results (Scalise et al., 2011)		
	Using the data for drawing tables and graphs (Scalise et al., 2011)		
	Data interpretation (Shin et al., 2003; Zacharia et al., 2008)		
	Forming conclusions from data (Ketelhut, 2007)		
	Theory revision (Rutten et al., 2012)		

(continued)

Table 3.3 (continued)

Elements of enquiry activities	Stages	Synthesis of stages	ICT possible
Conclusion	Drawing conclusions (Dori & Sasson, 2008) Build the conceptual understanding (Quellmalz et al., 2012) Using and developing models (Bell et al., 2010; Zacharia et al., 2008) Plotting the data back in the computer (Scalise et al., 2011)	Systematise and register	Wikis Simulation software Tool to draw graphs, tables, diagrams Mental maps tools
	Communicating the results (Barab et al., 2007; Bell et al., 2010; NRC, 1996; Shin et al., 2003; Zacharia et al., 2008) Asking questions (Barab et al., 2007; Jaakkola & Nurmi, 2008) Arguing theories (Zacharia et al., 2008) Defending a conclusion based on evidence (Ucar & Trundle, 2011) Communicating the hypothesis to others with a continual cycling back and forth among all the activities (Barab et al., 2007) Communicating and defending scientific arguments (Dori & Sasson, 2008; Kim et al., 2012)	Communicate the results	Chat Discussion forum Wiki (or Google Docs) Messaging system
	Prediction (Bell et al., 2010) Considering solutions in terms of their societal impacts (Barab et al., 2007) Critiquing the investigations of others (Quellmalz et al., 2012) Scaffolding needs for successful inquiry (Scalise et al., 2011)	Apply knowledge to new situations	Video Photo Wiki Discussion forum Simulation software

presented in Table 3.3. The five categories, called *elements of inquiry activities*, are accompanied by 12 inquiry-specific broad steps that constitute a "*synthesis of stages*" and are related to the *possible ICT* that can be employed.

(1) *Presenting the problem and reflecting on it* is almost always the first step in IBSE. Three new processes in this step were synthesised: (a) *Explore the world* – the students make observations and look at the scientific phenomena that are significant or arouse their curiosity. They can raise current topics of interest and socio-scientific and socially relevant questions. (b) *Present a problem* – the students or teacher presents a problem, which may be an open problem, a research problem or a problem situation. (c) *Reflect on the problem* – in this stage, it is crucial that the teachers clarify the real motive for designing, choosing and presenting the problem. It is important that they coordinate discussions with the whole class to encourage conceptions and

Fig. 3.1 The elements that characterise the steps involved in inquiry activities

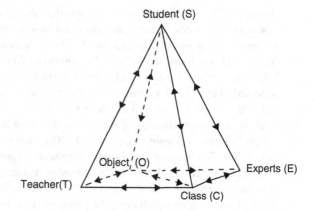

consolidate the processes experienced by the students during this stage. A particular difficulty is formulating "good" questions that are relevant and can be studied using a number of tools and scientific methods (Lucero et al., 2012). For this stage of the *inquiry*, many studies use *videos*, *Web*, *hypermedia*, *multimedia*, *MUVE*, *photos*, *simulations* and *simulation software*. The interactions (as shown in Fig. 3.1) can start with the teacher (T → S → O → T) or the student (S → O → C → T) or can be interactions among the student, teacher, object and class (S ↔ T ↔ O ↔ C).

(2) *Generating hypotheses* is the second step in IBSE (Barab et al., 2007; Bell et al., 2010; Dori & Sasson, 2008; Jaakkola & Nurmi, 2008; Ketelhut, 2007; Quellmalz et al., 2012; Rutten et al., 2012; Ucar & Trundle, 2011). For us, important stages for this step are frequently not taken into consideration: (a) formulating explanations for the problem presented; (b) demonstrating logical connections among scientific concepts to guide the development of a hypothesis and the design of an experiment (Ebenezer et al., 2011); (c) raising questions that guide the generation of hypotheses; and (d) writing down the hypotheses. The next step is to assess the validity of the hypotheses, i.e. gather evidence to assess their robustness (Quellmalz et al., 2012). When developing the hypotheses, the student may wish to use spreadsheets from *Google Docs* to facilitate teamwork and use social networks and messaging systems to share hypotheses. To test the hypotheses, Bell et al. (2010) suggest using simulation software, and this may be enacted using the cycle of interactions in Fig. 3.1: S → O → T → S and/or S → O → C → T, for example.

(3) The *investigative process* is the third step and can be organised in two stages: *planning* and *investigating*. For the first stage, some characteristics are important (Table 3.3): (a) planning a study to address the research question (Squire & Jan, 2007); (b) developing methods, tools and a rationale to explain the results (Squire & Jan, 2007); (c) developing theories based on evidence; (d) studying other research (Squire & Jan, 2007); and (e) designing experiments (Scalise et al., 2011). The second stage – investigating – can also be performed using books and other sources of information to collate what is known on the

subject (NRC, 1996). It is at this stage that data are collected and evidence assessed; discoveries occur; and data and results are obtained using diverse research instruments (Ebenezer et al., 2011). Various data collection instruments can be used, including the Internet, Web databanks, simulations, MUVEs, and software, for example. The interactions that occur may also be reflected in Fig. 3.1, including those that involve experts but mainly those that are of the forms S → O → S and S → O → C.

(4) *Analysis and interpretation of the data* collected is a fundamental competence for performing an *inquiry-based activity*. This step is characterised by two processes. The first is *analysing the data obtained*: (a) proposing answers, explanations and predictions (NRC, 1996); (b) using mathematical tools and statistical software to analyse and display data in tables and graphs (Ebenezer et al., 2011); (c) testing and evaluating the plausibility of discoveries in the search for new knowledge (Barab et al., 2007; Jaakkola & Nurmi, 2008); (d) using mathematics to analyse research and communicate information (Kim et al., 2012); and (e) sharing data to organise the information. After organising and analysing the data collected, the student should *interpret the information found* or, in other words, (a) be able to explain the process followed to arrive at the results and to evaluate the evidence using scientific criteria to arrive at results and formulate models (Ebenezer et al., 2011); (b) formulate, revise and analyse scientific explanations and models using logic and evidence (Barab et al., 2007; Kim et al., 2012); (c) justify the results (Quellmalz et al., 2012; Shin et al., 2003; Zydney & Grincewicz, 2011); (d) analyse errors and noisy data (Scalise et al., 2011); (e) synthesise the results (Scalise et al., 2011); (f) use the data in tables and graphs (Scalise et al., 2011); (g) draw conclusions from the data (Ketelhut, 2007); and (h) review the theory (Rutten et al., 2012). The instruments used for analysis include Excel spreadsheets, software for generating graphs and tables, simulations and wikis, for example. Exchanges among the student, virtual object and the class (as in Fig. 3.1) are fundamental for analysing the results in a collaborative and constructivist manner (Oshima et al., 2004).

(5) The last step is the *conclusion* of the activities, which is divided into three new processes: (a) *Systematising and registering the results*: these can be printed or recorded directly on the electronic device. At this stage, the students draw conclusions (Dori & Sasson, 2008), build a conceptual understanding (Quellmalz et al., 2012), use and develop models (Bell et al., 2010; Zacharia et al., 2008) and plot the data on the computer (Scalise et al., 2011). (b) *Communicating the results*: this stage relies strongly on scientific argumentation such that the students can communicate, explain and defend their results (Barab et al., 2007; Bell et al., 2010; Dori & Sasson, 2008; Kim et al., 2012; NRC, 1996; Shin et al., 2003; Zacharia et al., 2008); defend a conclusion based on evidence (Ucar & Trundle, 2011); and discuss theories (Zacharia et al., 2008) with other colleagues and the teacher(s). (c) *Applying the knowledge acquired to new situations*: this is a concluding stage in which the student can reflect on what has been achieved and use the results from the inquiry activity as a hypothesis for other activities. It is the moment at which connections are made (NRC, 2000),

thereby relating the activity with everyday life, analysing the results in the light of what the student know about the reality of things and predicting real phenomena (Bell et al., 2010). Some studies define this moment as considering the solutions in terms of their societal impacts (Barab et al., 2007), critiquing the investigations of others (Quellmalz et al., 2012) and scaffolding needs for successful inquiry (Scalise et al., 2011). Communication of the results can be direct, using scientific arguments or even supported by an *educational technology*, such as wikis, chats, e-mail and online forums, for example. The interaction of the form $S \rightarrow O \rightarrow C \rightarrow T$ is fundamental for concluding the activity. Once again, we note that these steps are not necessarily linear – the teacher, together with the students, can raise questions and construct new hypotheses during the development of an *inquiry activity*. Some studies indicate the existence of a wide choice of computer tools that are capable of helping students focus on higher learning processes, which are characteristic of *IBSE*. The studies also indicate that the computer is a mediating tool for the information that can be controlled by the students themselves and is able to support them in planning investigations and building knowledge. The necessary competencies for building knowledge and that use ICT are part of routine processes such as performing calculations on, collecting, classifying and visualising data. The students can access information and get help through digital interfaces when they wish and do not necessarily have to rely on the teacher, as shown in Fig. 3.1.

3.6 Conclusions

In recent decades, a variety of teaching strategies have been developed to aid the understanding of science subjects. In this context, we presented how *inquiry-based instruction* is a possible approach that has been gaining support in the curricula of various countries.

Hands-on IBSE is a teaching approach that is widely accepted, although there are various interpretations of what constitutes *inquiry-based instruction* (Bell et al., 2010; Jaakkola & Nurmi, 2008; Ucar & Trundle, 2011). There are also various interpretations of the effectiveness of the *inquiry-based approach* and how it should be implemented, regardless of the possible definitions (see Table 3.2).

In this chapter, we sought to extend the *inquiry-based approach* through the use of ICT because we found that students are especially motivated and engaged to take part in *inquiry-based instruction* when the learning is supported by some form of technology (Smetana & Bell, 2012; Squire & Jan, 2007; Stieff, 2011; Zacharia, 2005). Initially we assumed that IBSE is a method for teaching and learning science that is supported by inquiry activities, aimed at developing certain cognitive skills, and characterised by specific steps and procedures. These procedures are made up of three phases: (a) planning (before), (b) development (during) and (c) reflection (after). By analysing the possible steps of the inquiry activities that are supported by the ICT, we found that ICT increased the possibility of performing the inquiry process, gave the students new skills and had effects that *hands-on* type of activities do not allow, including sharing of data that are available on the Web, open and guided research, research strategies, synchronous and asynchronous communication, collaborative work and collaboration with scientists, for example.

Table 3.2 presents some examples that use IBSE as a theoretical framework, the use of educational technologies and the effects of these technologies. However, the design and use of a number of ICT tools as part of inquiry activities requires a certain degree of caution because an excess of resources can inhibit the learning process (Waight & Abd-El-Khalick, 2007). To minimise this problem, it is the responsibility of the teacher to manage the learning process, given that the technological tools allow them to be active participants in the learning process (Lee, Buxton, Lewis, & LeRoy, 2006). The teacher should pay attention to the progress of the inquiry activities, whose steps normally follow the cycle in Fig. 3.1, in which the activities are centred on the student, mediated by the teacher or object (that can be experimental or virtual), and require the object but not necessarily the teacher and the student can interact with the object, class, teacher and/or expert (thereby complementing the inquiry). By compiling a variety of approaches to IBSE, we created a table, which is ordered using a set of five categories that characterise the main ideas behind the IBSE stages included in Table 3.3. The categories, which are referred to as *elements of inquiry activities*, are accompanied by 12 inquiry-specific broad steps that constitute a *synthesis of stages* and related to the *possible ICT* that can be employed.

Table 3.3 were elaborated to cover a wide range of stages used in *inquiry activities* in science teaching and understand IBSE from an educational technology perspective. Table 3.3 do not provide a complete overview of all of the steps needed to characterise IBSE, the *inquiry activities* and the use of the educational technologies that are part of this type of teaching. Our intention was to present a theoretical framework that could characterise IBSE, present the results of IBSE that use ICT and define a structure for the possible steps to be used in the context of ICT. We still need to validate the *elements of inquiry activities* and the *synthesis of stages* and relate them with the *possible ICTs* listed in to understand *ICT-on and IBSE*.

References

AAAS. (1990). *Science for all Americans*. New York, NY: Oxford University Press.

Barab, S., Sadler, T., Heiselt, C., Hickey, D., & Zuiker, S. (2007). Relating narrative, inquiry, and inscriptions: Supporting consequential play. *Journal of Science Education and Technology, 16*, 59–82. https://doi.org/10.1007/s10956-006-9033-3

Barak, M., & Dori, Y. (2011). Science education in primary schools: Is an animation worth a thousand pictures? *Journal of Science Education and Technology, 20*, 608–620. https://doi.org/10.1007/s10956-011-9315-2

Bayraktar, S. (2001). A Meta-analysis of the Effectiveness of Computer-Assisted Instruction in Science Education. *Journal of Research on Technology in Education, 34*(2), 173–188. https://doi.org/10.1080/15391523.2001.10782344

Bell, R. L., & Trundle, K. C. (2008). The use of a computer simulation to promote scientific conceptions of moon phases. *Journal of Research in Science Teaching, 45*(3), 346–372. https://doi.org/10.1002/tea.20227

Bell, T., Urhahne, D., Schanze, S., & Ploetzner, R. (2010). Collaborative inquiry learning: Models, tools, and challenges. *International Journal of Science Education, 32*(3), 349–377. https://doi.org/10.1080/09500690802582241

Boaventura, D., Faria, C., Chagas, I., & Galvão, C. (2011). Promoting science outdoor activities for elementary school children: Contributions from a research laboratory. *International Journal of Science Education, 35*, 1–19. https://doi.org/10.1080/09500693.2011.583292

Bossler, A. P., Baptista, M., Freire, A. M. V., & do Nascimento, S. S. (2009). O estudo das vozes de alunos quando estão envolvidos em atividades de investigação em aulas de física. *Ensaio Pesquisa em Educação em Ciências, 11*(2), 307–319.

Brasil. (1999). Ministério da Educação. PCNEM – Parâmetros Curriculares Nacionais - Secretaria de Educação Média e Tecnológica. MEC; SEMTC.

Brasil. (2002). Ministério da Educação. PCN+ Ensino Médio: Orientações educacionais complementares aos Parâmetros Curriculares Nacionais - Ciências da Natureza e suas Tecnologias. Secretaria de Educação Média e Tecnológica: MEC; SEMTC.

Cachapuz, A., Gil-Pérez, D., de Carvalho, A. M. P., Praia, J., & Vilches, A. (2011). *A necessária renovação do ensino das ciências*. São Paulo, Brazil: Cortez Editora.

Charlier, B., Peraya, D., & Collectif. (2007). *Transformation des regards sur la recherche en technologie de l'éducation*. Bruxelles, Belgium: De Boeck.

Charpak, G. (1999). *Crianças investigadores e cidadãos*. Lisboa, Portugal: Instituto Piaget.

Chen, W., & Looi, C.-K. (2011). Active classroom participation in a group scribbles primary science classroom. *British Journal of Educational Technology, 42*, 676–686. https://doi.org/10.1111/j.1467-8535.2010.01082.x

Cher Ping, L. (2008). Global citizenship education, school curriculum and games: Learning mathematics, english and science as a global citizen. *Computers & Education, 51*, 1073–1093. https://doi.org/10.1016/j.compedu.2007.10.005

Clark, D., & Sampson, V. D. (2007). Personally-seeded discussions to scaffold online argumentation. *International Journal of Science Education, 29*, 253–277. https://doi.org/10.1080/09500690600560944

de Carvalho, A. M. P., & Collectif. (2004). *Ensino de ciências: Unindo a pesquisa e a prática*. São Paulo, Brazil: Pioneira Thomson Learning.

de Carvalho, A. M. P., & Collectif. (2013). *Ensino de Ciências por Investigação: Condições para implementação em sala de aula*. São Paulo, Brazil: Cengage Learning.

de Carvalho, A. M. P., Vannucchi, A. I., Barros, M. A., Gonçalves, M. E. R., & Rey, R. C. (2010). *Ciências no ensino fundamental: O conhecimento físico*. São Paulo, Brazil: Editora Scipione.

Dede, C., & Barab, S. (2009). Emerging technologies for learning science: A time of rapid advances. *Journal of Science Education and Technology, 18*, 301–304. https://doi.org/10.1007/s10956-009-9172-4

Dewey, J. (2007). *Logic - The theory of inquiry*. New York, NY: Saerchinger Press.

Donnelly, D., McGarr, O., & O'Reilly, J. (2011). A framework for teachers' integration of ICT into their classroom practice. *Computers & Education, 57*(2), 1469–1483. https://doi.org/10.1016/j.compedu.2011.02.014

Donnelly, D., O'Reilly, J., & McGarr, O. (2012). Enhancing the student experiment experience: Visible scientific inquiry through a virtual chemistry laboratory. *Research in Science Education, 43*(4), 1–22. https://doi.org/10.1007/s11165-012-9322-1

Dori, Y. J., & Sasson, I. (2008). Chemical understanding and graphing skills in an honors case-based computerized chemistry laboratory environment: The value of bidirectional visual and textual representations. *Journal of Research in Science Teaching, 45*, 219–250. https://doi.org/10.1002/tea.20197

Dori, Y. J., Tal, R. T., & Peled, Y. (2002). Characteristics of science teachers who incorporate web-based teaching. *Research in Science Education, 32*(4), 511–547. https://doi.org/10.102 3/A:1022499422042

Driver, R., Leach, J., Millar, R., & Scott, P. (1996). *Young people's images of science*. Buckingham, UK: Open University Press.

Driver, R., Newton, P., & Osborne, J. (2000). Establishing the norms of scientific argumentation in classrooms. *Science Education, 84*(3), 287–312. https://doi.org/10.1002/ (SICI)1098-237X(200005)84:3<287::AID-SCE1>3.0.CO;2-A

Ebenezer, J., Kaya, O. N., & Ebenezer, D. L. (2011). Engaging students in environmental research projects: Perceptions of fluency with innovative technologies and levels of scientific inquiry abilities. *Journal of Research in Science Teaching, 48*, 94–116. https://doi.org/10.1002/tea. 20387

Enyedy, N., Danish, J., Delacruz, G., & Kumar, M. (2012). Learning physics through play in an augmented reality environment. *International Journal of Computer-Supported Collaborative Learning, 7*, 347–378. https://doi.org/10.1007/s11412-012-9150-3

Furberg, A., & Ludvigsen, S. (2008). Students' meaning-making of socio-scientific issues in computer mediated settings: Exploring learning through interaction trajectories. *International Journal of Science Education, 30*, 1775–1799. https://doi.org/10.1080/09500690701543617

Galvão, C., Reis, P., Freire, S., & Almeida, P. (2011). Enhancing the popularity and the relevance of science teaching in Portuguese science classes. *Research in Science Education, 41*(5), 651–666. https://doi.org/10.1007/s11165-010-9184-3

Gelbart, H., Brill, G., & Yarden, A. (2009). The impact of a web-based research simulation in bioinformatics on students' understanding of genetics. *Research in Science Education, 39*, 725–751. https://doi.org/10.1007/s11165-008-9101-1

Gil Pérez, D., & Castro, P. V. (1996). La orientación de las prácticas de laboratorio como investigación: un ejemplo ilustrativo. *Enseñanza de Las Ciencias: Revista de Investigación Y Experiencias Didácticas, 14*(2), 155–164.

Hakkarainen, K. (2003). Progressive inquiry in a computer-supported biology class. *Journal of Research in Science Teaching, 40*, 1072–1088. https://doi.org/10.1002/tea.10121

Hansson, L., Redfors, A., & Rosberg, M. (2011). Students' socio-scientific reasoning in an astrobiological context during work with a digital learning environment. *Journal of Science Education and Technology, 20*, 388–402. https://doi.org/10.1007/s10956-010-9260-5

Hoffman, J. L., Wu, H.-K., Krajcik, J. S., & Soloway, E. (2003). The nature of middle school learners' science content understandings with the use of on-line resources. *Journal of Research in Science Teaching, 40*, 323–346. https://doi.org/10.1002/tea.10079

Jaakkola, T., & Nurmi, S. (2008). Fostering elementary school students' understanding of simple electricity by combining simulation and laboratory activities. *Journal of Computer Assisted Learning, 24*, 271–283. https://doi.org/10.1111/j.1365-2729.2007.00259.x

Jaakkola, T., Nurmi, S., & Veermans, K. (2011). A comparison of students' conceptual understanding of electric circuits in simulation only and simulation-laboratory contexts. *Journal of Research in Science Teaching, 48*, 71–93. https://doi.org/10.1002/tea.20386

Kawalkar, A., & Vijapurkar, J. (2013). Scaffolding science talk: The role of teachers' questions in the inquiry classroom. *International Journal of Science Education, 35*(12), 2004–2027. https:// doi.org/10.1080/09500693.2011.604684

Ketelhut, D. (2007). The impact of student self-efficacy on scientific inquiry skills: An exploratory investigation in river city, a multi-user virtual environment. *Journal of Science Education and Technology, 16*, 99–111. https://doi.org/10.1007/s10956-006-9038-y

Kim, H., & Herbert, B. (2012). Inquiry resources collection as a boundary object supporting meaningful collaboration in a wiki-based scientist-teacher community. *Journal of Science Education and Technology, 21*, 504–512. https://doi.org/10.1007/s10956-011-9342-z

Kim, H., Miller, H., Herbert, B., Pedersen, S., & Loving, C. (2012). Using a wiki in a scientist-teacher professional learning community: Impact on teacher perception changes. *Journal of Science Education and Technology, 21*(4), 440–452. https://doi.org/10.1007/s10956-011-9336-x

Kong, S. C., Yeung, Y. Y., & Wu, X. Q. (2009). An experience of teaching for learning by observation: Remote-controlled experiments on electrical circuits. *Computers & Education, 52,* 702–717. https://doi.org/10.1016/j.compedu.2008.11.011

Kubasko, D., Jones, M. G., Tretter, T., & Andre, T. (2008). Is it live or is it memorex? Students' synchronous and asynchronous communication with scientists. *International Journal of Science Education, 30,* 495–514. https://doi.org/10.1080/09500690701217220

Lee, H.-S., Linn, M. C., Varma, K., & Liu, O. L. (2010). How do technology-enhanced inquiry science units impact classroom learning? *Journal of Research in Science Teaching, 47*(1), 71–90. https://doi.org/10.1002/tea.20304

Lee, O., Buxton, C., Lewis, S., & LeRoy, K. (2006). Science inquiry and student diversity: Enhanced abilities and continuing difficulties after an instructional intervention. *Journal of Research in Science Teaching, 43*(7), 607–636. https://doi.org/10.1002/tea.20141

Lee, S. W., Tsai, C., Wu, Y., Tsai, M., Liu, T., Hwang, F., et al. (2011). Internet-based science learning: A review of journal publications. *International Journal of Science Education, 33,* 1893–1925. https://doi.org/10.1080/09500693.2010.536998

Lin, H., Hong, Z., Chen, C., & Chou, C. (2011). The effect of integrating aesthetic understanding in reflective inquiry activities. *International Journal of Science Education, 33*(9), 1199–1217. https://doi.org/10.1080/09500693.2010.504788

Lin, J. M., Wang, P., & Lin, I. (2012). Pedagogy*technology: A two-dimensional model for teachers' ICT integration. *British Journal of Educational Technology, 43*(1), 97–108. https://doi.org/10.1111/j.1467-8535.2010.01159.x

Lin, L.-F., Hsu, Y.-S., & Yeh, Y.-F. (2012). The role of computer simulation in an inquiry-based learning environment: Reconstructing geological events as geologists. *Journal of Science Education and Technology, 21,* 370–383. https://doi.org/10.1007/s10956-011-9330-3

Lindgren, R., & Schwartz, D. L. (2009). Spatial learning and computer simulations in science. *International Journal of Science Education, 31,* 419–438. https://doi.org/10.1080/09500 690802595813

Looi, C.-K., Zhang, B., Chen, W., Seow, P., Chia, G., Norris, C., & Soloway, E. (2011). 1:1 mobile inquiry learning experience for primary science students: A study of learning effectiveness. *Journal of Computer Assisted Learning, 27,* 269–287. https://doi.org/10.1111/j.1365-2729.2010.00390.x

Lowe, D., Newcombe, P., & Stumpers, B. (2012). Evaluation of the use of remote laboratories for secondary school science education. *Research in Science Education, 43,* 1–23. https://doi.org/10.1007/s11165-012-9304-3

Lucero, M., Valcke, M., & Schellens, T. (2012). Teachers' beliefs and self-reported use of inquiry in science education in public primary schools. *International Journal of Science Education, 35,* 1–17. https://doi.org/10.1080/09500693.2012.704430

Mayer, R. E. (2009). *Teoria cognitiva da aprendizagem multimédia. In Ensino online e aprendizagem multimédia.* Lisboa, Portugal: Relógio D'Água Editores.

Ministère de l'Éducation et du développement de la petite enfance. (2011). *Programme d'études: Sciences et technologies 6e anné - 8e année.* New Nouveau Brunswick.

Mistler-Jackson, M., & Butler Songer, N. (2000). Student motivation and internet technology: Are students empowered to learn science? *Journal of Research in Science Teaching, 37,* 459–479. https://doi.org/10.1002/(SICI)1098-2736(200005)37:5<459::AID-TEA5>3.0.CO;2-C

Nelson, B. (2007). Exploring the use of individualized, reflective guidance in an educational multi-user virtual environment. *Journal of Science Education and Technology, 16,* 83–97. https://doi.org/10.1007/s10956-006-9039-x

NGSS Lead States (2013). *Next Generation Science Standards: For States, By States.* Washington, DC: The National Academies Press. https://doi.org/10.17226/18290

NRC. (1996). *National science education standards*. Washington, DC: National Academy Press. Retrieved from http://www.nap.edu/catalog.php?record_id=4962

NRC. (2000). *Inquiry and the national science education standards: A guide for teaching and learning*. Washington, DC: National Academy Press. Retrieved from http://www.nap.edu/catalog.php?record_id=9596

NRC. (2012). *A framework for K-12 science education: Practices, crosscutting concepts, and core ideas*. Washington, DC: The National Academies Press. Retrieved from http://www.nap.edu/catalog/13165/a-framework-for-k-12-science-education-practices-crosscutting-concepts

Oh, P. S. (2010). How can teachers help students formulate scientific hypotheses? Some strategies found in abductive inquiry activities of earth science. *International Journal of Science Education, 32*(4), 541–560. https://doi.org/10.1080/09500690903104457

Olympiou, G., & Zacharia, Z. C. (2012). Blending physical and virtual manipulatives: An effort to improve students' conceptual understanding through science laboratory experimentation. *Science Education, 96*, 21–47. https://doi.org/10.1002/sce.20463

Oshima, J., Oshima, R., Murayama, I., Inagaki, S., Takenaka, M., Nakayama, H., & Yamaguchi, E. (2004). Design experiments in Japanese elementary science education with computer support for collaborative learning: Hypothesis testing and collaborative construction. *International Journal of Science Education, 26*, 1199–1221. https://doi.org/10.1080/0950069032000138824

Posner, G. J., Strike, K. A., Hewson, P. W., & Gertzog, W. A. (1982). Accommodation of a scientific conception: Toward a theory of conceptual change. *Science Education, 66*(2), 211–227. https://doi.org/10.1002/sce.3730660207

Pyatt, K., & Sims, R. (2012). Virtual and physical experimentation in inquiry-based science labs: Attitudes, performance and access. *Journal of Science Education and Technology, 21*, 133–147. https://doi.org/10.1007/s10956-011-9291-6

Quellmalz, E. S., Timms, M. J., Silberglitt, M. D., & Buckley, B. C. (2012). Science assessments for all: Integrating science simulations into balanced state science assessment systems. *Journal of Research in Science Teaching, 49*, 363–393. https://doi.org/10.1002/tea.21005

Russell, D. W., Lucas, K. B., & McRobbie, C. J. (2004). Role of the microcomputer-based laboratory display in supporting the construction of new understandings in thermal physics. *Journal of Research in Science Teaching, 41*, 165–185. https://doi.org/10.1002/tea.10129

Rutten, N., van Joolingen, W. R., & van der Veen, J. T. (2012). The learning effects of computer simulations in science education. *Computers & Education, 58*, 136–153. https://doi.org/10.1016/j.compedu.2011.07.017

Sasseron, L. H., & Carvalho, A. M. (2011). Scientific literacy: A bibliographical review. *Investigações Em Ensino de Ciências, 16*(1), 59–77.

Scalise, K., Timms, M., Moorjani, A., Clark, L., Holtermann, K., & Irvin, P. S. (2011). Student learning in science simulations: Design features that promote learning gains. *Journal of Research in Science Teaching, 48*, 1050–1078. https://doi.org/10.1002/tea.20437

Schiel, D., Orlandi, A. S., & Collectif. (2009). *Ensino de ciências por investigação*. São Carlos, Brazil: CDCC/Compacta Gráfica e Editora Ltda.

She, H.-C., & Chen, Y.-Z. (2009). The impact of multimedia effect on science learning: Evidence from eye movements. *Computers & Education, 53*, 1297–1307. https://doi.org/10.1016/j.compedu.2009.06.012

Shin, N., Jonassen, D. H., & McGee, S. (2003). Predictors of well-structured and ill-structured problem solving in an astronomy simulation. *Journal of Research in Science Teaching, 40*, 6–33. https://doi.org/10.1002/tea.10058

Smetana, L. K., & Bell, R. L. (2012). Computer simulations to support science instruction and learning: A critical review of the literature. *International Journal of Science Education, 34*, 1337–1370. https://doi.org/10.1080/09500693.2011.605182

Snir, J., Smith, C. L., & Raz, G. (2003). Linking phenomena with competing underlying models: A software tool for introducing students to the particulate model of matter. *Science Education, 87*(6), 794–830. https://doi.org/10.1002/sce.10069

So, W. W. M. (2012). Creating a framework of a resource-based e-learning environment for science learning in primary classrooms. *Technology, Pedagogy and Education, 21*(3), 317–335. https://doi.org/10.1080/1475939X.2012.719399

Songer, N. B., Lee, H.-S., & Kam, R. (2002). Technology-rich inquiry science in urban classrooms: What are the barriers to inquiry pedagogy? *Journal of Research in Science Teaching, 39*(2), 128–150. https://doi.org/10.1002/tea.10013

Soong, B., & Mercer, N. (2011). Improving students' revision of physics concepts through ICT-based co-construction and prescriptive tutoring. *International Journal of Science Education, 33*(8), 1055–1078. https://doi.org/10.1080/09500693.2010.489586

Squire, K., & Jan, M. (2007). Mad city mystery: Developing scientific argumentation skills with a place-based augmented reality game on handheld computers. *Journal of Science Education and Technology, 16*, 5–29. https://doi.org/10.1007/s10956-006-9037-z

Stieff, M. (2011). Improving representational competence using molecular simulations embedded in inquiry activities. *Journal of Research in Science Teaching, 48*, 1137–1158. https://doi.org/10.1002/tea.20438

Tan, A.-L., & Wong, H.-M. (2012). "Didn't get expected answer, rectify it": Teaching science content in an elementary science classroom using hands-on activities. *International Journal of Science Education, 34*(2), 197–222. https://doi.org/10.1080/09500693.2011.565378

Tolentino, L., Birchfield, D., Megowan-Romanowicz, C., Johnson-Glenberg, M. C., Kelliher, A., & Martinez, C. (2009). Teaching and learning in the mixed-reality science classroom. *Journal of Science Education and Technology, 18*, 501–517. https://doi.org/10.1007/s10956-009-9166-2

Trópia, G. (2011). Percursos históricos de ensinar ciências através de atividades investigativas. *Ensaio Pesquisa em Educação em Ciências, 13*(1), 121.

Ucar, S., & Trundle, K. C. (2011). Conducting guided inquiry in science classes using authentic, archived, web-based data. *Computers & Education, 57*(2), 1571–1582. https://doi.org/10.1016/j.compedu.2011.02.007

Van Zee, E., & Roberts, D. (2006). Making science teaching and learning visible through web-based "snapshots of practice". *Journal of Science Teacher Education, 17*(4), 367–388. https://doi.org/10.1007/s10972-006-9027-2

Varma, K., & Linn, M. (2012). Using interactive technology to support students' understanding of the greenhouse effect and global warming. *Journal of Science Education and Technology, 21*(4), 453–464. https://doi.org/10.1007/s10956-011-9337-9

Waight, N., & Abd-El-Khalick, F. (2007). The impact of technology on the enactment of "inquiry" in a technology enthusiast's sixth grade science classroom. *Journal of Research in Science Teaching, 44*, 154–182. https://doi.org/10.1002/tea.20158

Waight, N., & Abd-El-Khalick, F. (2011). From scientific practice to high school science classrooms: Transfer of scientific technologies and realizations of authentic inquiry. *Journal of Research in Science Teaching, 48*(1), 37–70. https://doi.org/10.1002/tea.20393

Wang, C., Ke, Y.-T., Wu, J.-T., & Hsu, W.-H. (2012). Collaborative action research on technology integration for science learning. *Journal of Science Education and Technology, 21*, 125–132. https://doi.org/10.1007/s10956-011-9289-0

Webb, M. E. (2005). Affordances of ICT in science learning: Implications for an integrated pedagogy. *International Journal of Science Education, 27*, 705–735. https://doi.org/10.1080/09500690500038520

Zacharia, Z. (2003). Beliefs, attitudes, and intentions of science teachers regarding the educational use of computer simulations and inquiry-based experiments in physics. *Journal of Research in Science Teaching, 40*(8), 792–823. https://doi.org/10.1002/tea.10112

Zacharia, Z. C. (2005). The impact of interactive computer simulations on the nature and quality of postgraduate science teachers' explanations in physics. *International Journal of Science Education, 27*(14), 1741–1767. https://doi.org/10.1080/09500690500239664

Zacharia, Z. C. (2007). Comparing and combining real and virtual experimentation: An effort to enhance students' conceptual understanding of electric circuits. *Journal of Computer Assisted Learning, 23*, 120–132. https://doi.org/10.1111/j.1365-2729.2006.00215.x

Zacharia, Z. C., Olympiou, G., & Papaevripidou, M. (2008). Effects of experimenting with physical and virtual manipulatives on students' conceptual understanding in heat and temperature. *Journal of Research in Science Teaching, 45*, 1021–1035. https://doi.org/10.1002/tea.20260

Zhang, B., Looi, C.-K., Seow, P., Chia, G., Wong, L.-H., Chen, W., et al. (2010). Deconstructing and reconstructing: Transforming primary science learning via a mobilized curriculum. *Computers & Education, 55*, 1504–1523. https://doi.org/10.1016/j.compedu.2010.06.016

Zompero, A. F., & Laburú, C. E. (2011). Atividades investigativas no ensino de ciências: Aspetos históricos e diferentes abordagens. *Ensaio Pesquisa em Educação em Ciências, 13*(3), 67.

Zydney, J., & Grincewicz, A. (2011). The use of video cases in a multimedia learning environment for facilitating high school students' inquiry into a problem from varying perspectives. *Journal of Science Education and Technology, 20*, 715–728. https://doi.org/10.1007/s10956-010-9264-1

Index

© The Author(s), under exclusive license to Springer Nature Switzerland AG 2019   93
G. W. Rocha Fernandes et al., *Using ICT in Inquiry-Based Science Education*,
SpringerBriefs in Education, https://doi.org/10.1007/978-3-030-17895-6

Printed in the United States
By Bookmasters